THE
SHADOW
WAR

THE SHADOW WAR

Inside Russia's and China's Secret
Operations to Defeat America

JIM SCIUTTO

HARPER
An Imprint of HarperCollinsPublishers

HarperCollins books may be purchased for educational, business, or sales promotional use. For information, please email the Special Markets Department at SPsales@harpercollins.com.

FIRST EDITION

Designed by William Ruoto

Library of Congress Cataloging-in-Publication Data has been applied for.

ISBN 978-0-06-285364-6

19 20 21 22 23   LSC   10 9 8 7 6 5 4 3 2 1

To Gloria, Tristan, Caden, and Sinclair
In memory of my parents, Ernest and Elizabeth Sciutto

CONTENTS

THE
SHADOW
WAR

CHAPTER 1

Inside the Shadow War

The senior government official went quiet whenever the waiter approached the table, restarting only once he had walked away. In the past, he had been one of the most difficult sources for me to meet. I would always be the one initiating contact and, most of the time, I got a no, or no answer at all. This time, however, he had requested the meeting. He was very careful by nature, so I knew he had something to say. His choice for lunch was an odd one for a private conversation. Café Milano is a caricature of the high-powered Washington eatery: overpriced food, obsequious staff, an expensive wine list, and a clientele comprising a who's who of Washington and international power brokers. And yet here we were discussing what was arguably Russia's boldest and most frightening overseas operation since the Cold War.

My source told me Western intelligence was now highly confident that Vladimir Putin himself had ordered and directed the poisoning of the former KGB agent Sergei Skripal and his daughter, Yulia, in Salisbury, England, earlier that spring. The attempted murder with the powerful Russian-made nerve agent Novichok had shocked the United Kingdom and Europe. The use of Novichok was particularly alarming. Many times more deadly than even the most powerful nerve agent ever in the US arsenal, VX (which has been banned for decades), Novichok kills by disrupting nerve signals throughout the body, causing repeated and uncontrollable muscle contractions. Victims are left convulsing in pain, vomiting, and foaming at the mouth, which is how witnesses found the Skripals that day on a park bench in Salisbury. In the months following the attack, every European official I met described it in frightening terms. By carrying out a potentially deadly operation on the soil of a NATO ally, they argued, Russia had set a new, frightening standard for its malign activities abroad.

US and Western intelligence agencies had very quickly surmised that such an operation could not have happened without the knowledge of senior Russian leaders, and in the top-heavy Kremlin, Putin was the only senior leader who mattered. However, a direct order from the Russian president to assassinate someone on British soil would elevate the stakes. And my source told me that Western intelligence agencies had now concluded it was "highly likely" Putin had done just that. With the Skripal operation, Putin appeared to have sent two bold messages: to the British and the West, that he saw no territorial limits to Russia's violent actions abroad; and to Russian dissidents and other critics, that they were not safe anywhere in the world.

My contact leaned in to share one more disturbing detail. British investigators had now determined that the two Russian military intelligence agents who had carried out the operation had brought enough Novichok into Britain to kill thousands.

"Thousands?" I asked to be certain.

"Yes, thousands," he repeated.

Western intelligence did not believe the Russian hit team intended to kill thousands of its citizens. However, the gall of transporting an extremely dangerous substance in such an immense quantity into the United Kingdom stunned Western leaders. Moving even a small amount of Novichok carried enormous risks to anyone who came into contact with it. That had been made clear when two Salisbury residents unconnected to the Skripals—Charlie Rowley and Dawn Sturgess—came across a vial of the substance innocently disguised as a bottle of Nina Ricci perfume, apparently discarded by the Skripals' attempted assassins. After spraying the substance on her wrist believing it was perfume, Sturgess fell ill within minutes and died days later; Rowley only narrowly survived. Remarkably, Skripal and his daughter survived as well, but only after weeks in the hospital. Smuggling immense quantities of Novichok into the United Kingdom obviously increased the risk of more casualties. Moscow seemed to be demonstrating how little it cared and, crucially, how little it feared Britain's and the West's response. This was a massive chemical weapons attack inside the West by Russia. It was unprecedented. Or was it?

As shocking as the Skripal poisoning was proving, each detail sounded remarkably familiar to me. Twelve years earlier, while based in London as senior foreign correspondent for ABC News, I had covered the assassination of the Russian dissident Alexander Litvinenko. In a plot seemingly stolen from the pages of a John le Carré novel, two Russian agents had poisoned Litvinenko with radioactive polonium-210 injected into his cup of tea. Just a speck of the substance was powerful enough to kill several people. And its radioactivity is so strong that British investigators were later able to trace its entire path into and around Britain, from seats 26E and 26F on the Russian jet the two agents had flown into London, to their room in a Best Western hotel on Shaftesbury Avenue in Piccadilly, to the Arsenal football stadium, where they had taken in a match, to the Itsu sushi restaurant where they first met Litvinenko, all the way to the Pine Bar of the Millennium Hotel in Mayfair, where they delivered the deadly dose.

The target and the weapon were different from the poisoning of Sergei Skripal, but the pattern was the same: an extraterritorial assassination—this one, successful—of a man the Kremlin viewed as an enemy of the state. Like Skripal, Litvinenko was a former agent of the FSB, the successor to the KGB. He had been expelled from the FSB in 1998 after making public allegations of illegal activity by the Russian intelligence services. His most explosive charge was contained in a book claiming the Russian president had staged a series of deadly terror attacks on Moscow apartment buildings in 1999, not the Chechen terrorists whom the Kremlin had blamed. The goal: to assure Putin's election in 2000 and provide justification for Russia's second military intervention in Chechnya.

In 2000, Litvinenko fled Russia with his wife and son to the United Kingdom, where he requested political asylum. A year later, he was granted asylum and went on to become a British citizen. In his new home in the West, in a NATO ally no less, he thought he would be safe. And he continued his work exposing what he claimed were the crimes of the Russian leadership. He allied himself with another London-based Russian dissident and Putin critic, Boris Berezovsky. Soon before his death, Litvinenko had accused Putin of ordering the murder earlier in 2006 of the Russian journalist Anna Politkovskaya. In the end, he like Skripal was still within reach of the FSB.

The 2006 operation was uniquely bold. The hotel where Litvinenko was poisoned—the Millennium—was just a half block from the US embassy in London. More alarmingly, the weapon was extremely powerful. At the time, alarmed British officials described it to me as the country's first-ever chemical weapons attack, comparing it to detonating a dirty bomb on the streets of London. And as with the Skripal poisoning, Russian operatives had put thousands of people in danger.

"Many thousands of members of the public, including British residents and visitors from overseas, might have been at risk from radioactivity," an investigating lawyer told the official British inquiry into the poisoning in 2016.

In the wake of the attack, British authorities would test some eight hundred people for contamination. Dozens were found to have elevated doses of radiation. Some, such as Litvinenko's wife and son, were contaminated by coming into direct contact with him. From those and other contaminated sites the radiation spread like the outbreak of a deadly pathogen. Others who had even passing contact with his family were also found to be contaminated. In turn, people who had come into passing contact with those secondary victims were found to be contaminated as well. The web of primary, secondary, tertiary contacts, and so on, would grow to hundreds.

In the midst of covering the story, I became a potential victim as well. Since in my reporting I had visited many of the sites where Litvinenko was believed to have been exposed to polonium-210, including the Itsu sushi restaurant and the Millennium Hotel, ABC News sent me for radiation tests. The details of the process are crass but it involved drinking a dye, lots of water, and then submitting gallons of urine samples in sufficient quantity to detect radioactive contamination. It was a nervous few days for me and my wife, just months after we had been married. Thankfully, my samples tested negative.

Still, in his closing remarks to the British inquiry, a lawyer representing the London police described the plot as "a nuclear attack on the streets of London."

"Anyone who arranges for polonium-210 to be brought into a city center does so without any regard for human life," Richard Horwell testified. "We will never know how dangerous the exposure of polonium to the public at large will be and what long-term effects will be visited upon Londoners."[1]

Polonium-210 is a lousy murder weapon for a murder you want to cover up. A nuclear expert who testified to the British inquiry would trace its origin to a particular Russian nuclear facility in the city of Sarov, several hundred miles south of Moscow. Investigators found traces of it everywhere the two suspects had gone, providing an indelible web of radioactive fingerprints. The highest concentrations were found at the table in the Millennium Hotel's Pine Bar where Litvinenko and his alleged killers, Andrei Lugovoy and Dmitry Kovtun, had met for tea, and inside the teapot they had been served with.

Yet, despite the overwhelming evidence, it would take Britain a full decade to officially blame Russia for the poisoning. A 2016 public inquiry concluded what Western intelligence had assessed within weeks of the attack: that Russia had carried out Litvinenko's murder, directing two operatives, one of whom was a former KGB bodyguard, to poison him with polonium-210 sourced from a Russian nuclear reactor. As with Skripal, the operation, the inquiry found, was likely ordered by Putin himself.

Sir Robert Owen, who led the inquiry, concluded, "I am sure that Mr. Lugovoy and Mr. Kovtun placed the polonium 210 in the teapot at the Pine Bar on 1 November 2006. I am also sure that they did this with the intention of poisoning Mr. Litvinenko."[2]

In 2006, twelve years before Skripal's poisoning alarmed the world, the Kremlin had already calculated it could get away with murder on Western soil. And it would be proved mostly correct. Britain's belated response was to expel four Russian diplomats, a full decade after Litvinenko died. In 2017, Congress would impose sanctions under the Magnitsky Act on Lugovoy, the only Russian national to be targeted by the United States. The penalties for the 2006 operation—delicately measured and long delayed—were clearly insufficient to change Russian behavior, perhaps laying the groundwork for a repeat on the streets of Salisbury in 2018. To add insult to grievous injury, Lugovoy would be elected a member of the Russian state Duma, where he still serves today.

Two deadly operations on Western soil, using weapons that threatened the lives of thousands, carried out under orders from the Russian president, twelve years apart. For Russia it is difficult to identify one single attack as the opening battle of its Shadow War on the United States and the West. However, the events of the last decade showed two consistent and disturbing lines: growing Russian aggression and persistent Western delusions about Russian intentions. The same pattern is discernible regarding China, which was launching its own inaugural battles in another, arguably more existentially dangerous Shadow War on the United States.

For Russia, the months that followed Litvinenko's murder brought a series of hostile acts of increasing boldness: its 2007 cyberattack on Estonia, its 2008 invasion of Georgia. In February 2014, Russia invaded and then annexed Crimea, Ukraine, slicing off a piece of a sovereign European nation without firing a shot. Soon after, it launched a war in Eastern Ukraine, arming "volunteers" to fight the Ukrainian armed forces and further destabilize the country. In cyberspace, from 2014 to 2015, Russia carried out a lengthy and expansive attack on the email system of the US Department of State—an operation that officials in the National Security Agency would later identify as a precursor to cyberattacks targeting the 2016 US presidential election. Russia's interference in 2016 carried its hostile activity to a new level of aggression, this one often described as a surprise attack on American democracy—a "political Pearl Harbor" that came without warning and therefore, understandably, caught the US national security community off guard. But, in fact, there were numerous warning signs prior to 2016 of a new and aggressive Russian strategy to undermine the United States at every turn, with a combination of hard and soft power.

China, America's other chief competitor internationally, was pursuing a similar strategy, with perhaps more subtlety but no less aggression. By the mid-2000s, China's national effort to steal US technology and state secrets was already in high gear and logging up enormous successes in both the public and private sectors. In 2014 China defied both international law and the laws of physics to manufacture entirely new sovereign territory in the middle of the South China Sea, beginning construction of a string of man-made islands in waters claimed by several of its Southeast Asian neighbors. China was also expanding its military capabilities and military footprint from under the waves all the way into space, with the express intention of surpassing the United States and—if necessary—defeating the United States in war.

Inside the US government and the intelligence community, these aggressive steps were first missed and then downplayed. US officials, led by President Barack Obama, accepted China's assurances it would not militarize its man-made islands in the South China Sea—assurances Beijing reneged on almost immediately. Obama would later accept Chinese assurances that Beijing would dial back its cyber theft of US corporate secrets, malicious activity that remains rampant and aggressive today. Even after finally acknowledging these acts of aggression, many US officials and policy experts continued to portray them as short-term or easily reversible.

With Russia, successive US leaders persisted in their conviction they could get the Russia relationship right, where their predecessors had failed. The Obama administration's ill-fated "reset" with Russia followed the Russian invasion of Georgia by just months. The image of then–secretary of state Hillary Clinton presenting her Russia counterpart Foreign Minister Sergei Lavrov with a mock-up red "reset button" in Geneva will long survive as a symbol of the West's chronic misreading of Moscow. Russian hackers had free rein inside the State Department's email network for months before they were detected. Later, not a single US intelligence agency predicted Russia's annexation of Crimea.

The Obama administration's dismissive view of the Kremlin would continue almost to the end of his administration. At the G7 summit in 2014, President Obama relegated Russia to "regional power" status, saying that its territorial ambitions "belonged in the nineteenth century." His 2014 comments echoed his disdain for Mitt Romney's foreign policy priorities in their October 2012 presidential debate: "When you were asked what was the biggest geopolitical threat facing America, you said Russia, not Al-Qaeda. You said Russia, and the 1980s are now calling to ask for their foreign policy back because the Cold War's been over for twenty years."

Romney's answer to Obama then now looks prescient. "Russia indicated it is a geopolitical foe," he said. "I'm not going to wear rose-colored glasses when it comes to Russia or Mr. Putin."

However, in 2016, Obama's dismissiveness would be replaced by President Donald Trump's own rose-colored view of Moscow and Putin. If the run-up to 2016 was riven by missed warning signs and halting responses, with the US response to Russia's interference in the 2016 presidential election, the United States risked moving from misguided inaction to willful negligence.

At the core of these repeated errors by administrations of both parties was a fundamental misreading of Russian and Chinese goals and intentions, colored by the hope—ultimately a false one—that Russia and China want what the United States wants.

"I met Vladimir Putin back in the 1990s," said Ashton Carter, who served as secretary of defense from 2015 to 2017, as well as a defense official in the 1990s. "It was clear to me, but I won't say it was clear to everyone in the defense or certainly the strategic community, that Vladimir Putin . . . set himself the objective of thwarting the West per se. And that was an almost insuperable barrier to dealing with him in a constructive way."

Carter says the predominant US government view of China suffered from a similar case of mirroring.

"China, whom I think in the 1990s we thought at least might take a path towards greater involvement in participating and reinforcing the security system that the United States largely created and that they benefited from," said Carter, "instead would take a turn towards the assertion by the Middle Kingdom of its place in the sun."

Beyond the fundamental misjudgment of US adversaries' mindset, there was a failure to recognize a fundamental change in what Russia and China were willing to do to accomplish their goals—and how they would do so. In effect, America's principal adversaries conceived of—and then waged—an entirely new kind of warfare on the United States and the West.

Today, the most senior US national security officials, who led the US national security establishment as this new threat materialized, acknowledge they failed to understand the depth and breadth of what they now identify as America's most pressing national security threat.

"We need to study harder how they do it because we don't naturally do it ourselves," General Michael Hayden, director of the CIA from 2006 to 2009, told me. "I know about aerial combat. I know about second- and third-echelon attacks, because we do that, but we don't do this."

"This" is hybrid warfare, in short, a strategy of attacking an adversary while remaining just below the threshold of conventional war—what military commanders and strategists refer to as the "gray zone"—using a range of hard- and soft-power tactics: from cyberattacks on critical infrastructure, to deploying threats to space assets, to information operations designed to spark domestic division, to territorial acquisition just short of a formal invasion. This is warfare conducted in the shadows—a Shadow War—though with consequences as concrete and lasting as those of all-out war.

———————

This is a book about what happens when the enemies of the West realize that while they are unlikely to win a shooting war, they have another path to victory. The United States and the West have had a tendency to misread what their enemies are doing, to see their actions through old lenses. They often get Russian and Chinese motives wrong, their goals wrong, and the long-term consequences wrong. Moreover, what the United States and the West see as their greatest strengths—open societies, military innovation, dominance of technology on earth and in space, long-standing leadership in global institutions—these countries are undermining or turning into weaknesses.

The United States needs a new guide to international conflict, because the old one isn't working. It's as if China and Russia have started a new Cold War and America didn't notice. The tactics are new and changing, but the goals have not changed. They want to become more powerful on the world stage by weakening and destabilizing the West, its allies, and the systems they depend on. These two adversaries are also showing other countries the way, with Iran and North Korea starting down the same road. And it's more than America in the crosshairs: they see every nation that is not helping them as a potential target.

Eventually the United States will come to think of this Shadow War as America's primary foreign policy problem, even though most American citizens currently know nothing about it. The sooner it becomes the focus of political debates and international meetings, the brighter—and safer—America's future will be.

The Shadow War is not the result of a secret plan, hidden deep in the recesses of Russian and Chinese intelligence services. Both the tactics and the thinking behind it have been hiding in plain sight. In February 2013, General Valery Gerasimov, chief of staff for the Russian Federation's military, laid out his country's strategy in detail in an essay published for the world to see in the weekly newsletter *Military-Industrial Kurier.*

"In the twenty-first century, we have seen a tendency toward blurring the lines between the states of war and peace," Gerasimov wrote in an article innocuously entitled "The Value of Science Is in the Foresight." "Wars are no longer declared and, having begun, proceed according to an unfamiliar template."[3]

Though Gerasimov was ostensibly describing how Russia believed its adversaries were conducting warfare in the modern age, his essay articulated with remarkable candor Russia's own strategy for waging war on its adversaries, principally the United States and the West, forming the basis of what Western intelligence officials now regularly refer to as the "Gerasimov Doctrine," encompassing both military and nonmilitary methods.

"The very 'rules of war' have changed," he wrote. "The role of non-military means of achieving political and strategic goals has grown, and, in many cases, they have exceeded the power . . . of weapons in their effectiveness."[4]

For a senior Russian commander outlining his country's military strategy in a public forum, Gerasimov was remarkably specific, identifying the exact tactics that Russia would employ the very next year in Crimea and Eastern Ukraine, including special forces posing as something other than soldiers of the Russian Federation.

"The open use of forces—often under the guise of peacekeeping and crisis regulation—is resorted to only at a certain stage, primarily for the achievement of final success in the conflict," Gerasimov wrote.

These were the "little green men" who would show up on the streets of Crimea, ostensibly, at the request of ethnic Russians fearing attacks from their fellow ethnic Ukrainian citizens. Today General Hayden sees the Gerasimov essay, in all its bluntness and clarity, as one of the most obvious missed warning signs.

"This was an attack against an unappreciated weakness from an unexpected direction," Hayden told me. "Unexpected, because we're thinking about it like this, while Gerasimov—even though he wrote it down, we've never read it—is thinking about it that way."

China's hybrid warfare doctrine—its strategy for winning in the gray zone—goes by a different name: "winning without fighting," or what the 2017 US National Security Strategy describes as "continuing competition," with the two sides neither fully at peace nor at war. Its man-made islands in the South China Sea are examples of this strategy in action. Like Russia in Crimea, China was able to secure sovereign territory in disputed waters without firing a shot.

However, US officials with direct experience confronting Chinese intelligence warn that Beijing does not shy away from conflict and violence it deems necessary. Bob Anderson led the FBI's counterintelligence division until 2015 and identified and captured dozens of Chinese spies operating inside the United States.

"The Chinese are as vicious or more vicious than the Russians," Anderson told me. "They will kill people at the drop of a hat. They will kill families at the drop of a hat. They will do it much more quietly inside of China or in one of their territories, but they absolutely will, if they have to.

"The Chinese are a very vicious intelligence culture," he added.

Today, emboldened by their successes, Russia and China are waging hybrid war against a whole host of adversaries, big and small. Former defense secretary Carter sees Russia's hybrid war in action across the entire length of Russia's border with Europe, including numerous NATO allies.

"It actually continues along all of their western coast with Europe," said Carter. "Trying to undermine and peel countries away and intimidate them by the planning and in some cases conducting operations where you try to make what is happening susceptible enough to a big lie."

On every front, "the big lie" is an essential part of the strategy. With the invasion of Crimea and Ukraine, that meant denying that what were obviously Russian troops were indeed Russian troops. With meddling in the 2016 US presidential election, that meant spreading fake news via Russian news outlets and social media to sow doubts about Russia's role and to amplify those US politicians who echo those doubts, including President Donald Trump himself.

"Putin is one of the experts at the big lie, where you do something, and you deny it and create enough uncertainty so at least the Russian people don't believe you're doing what you're doing," Carter said.

In the case of Russia's election interference, some Americans believed the big lie as well, led by a US presidential candidate, then president, whose rhetoric often mimicked Russia's, sometimes word for word.

"They grab American-created memes for their social media attacks, generally from the alt-right, occasionally from the president," said General Hayden.

China conducts its own information operations, including through the growing international presence of its state-run media. In late 2016, China rebranded the international wing of its state-run China Central Television (CCTV) network as the China Global Television Network (CGTN), which maintains as broad a presence inside the United States as Russia's RT Network, but with less public awareness of its government backing. In CGTN's coverage, often delivered by American reporters and anchors, China's man-made islands are not a land grab but a question of sovereignty, challenged under treaties that, CGTN notes correctly, even the United States has not ratified.

The Shadow War has been years in the making. But China and Russia were firing the first shots when the United States was fixated on another threat and another kind of warfare—in the Mideast in the years following the September 11 attacks.

"About the time of the launch of the first Iraq War," said Carter, "both of these turns on the parts of Russia and China strategically had already occurred. But that is the very moment at which we entered what was really more than a decade of preoccupation elsewhere.

"I think that during that period, which spanned obviously two administrations, there was just a disinclination to face the fact that two major additional headaches were developing at the same time we were struggling. And our military was so preoccupied with the ones in front of us, which were terrorism and counterinsurgency in both Afghanistan and Iraq," said Carter.

Today, belatedly, hybrid warfare and the means of defending against it and winning in gray zone conflicts have moved to the top of the minds of US military commanders and intelligence officials. Starting in 2015, NATO developed a new war plan for defending Europe from Russian aggression—a plan that for the first time identifies and even incorporates hybrid warfare tactics.

"We hadn't had a war plan for twenty-five years," said Carter. "We hadn't thought we needed one."

That thinking does not yet hold sway in the White House. US defense and intelligence officials serving both the Obama and Trump administrations have told me repeatedly that the United States cannot effectively and credibly defend against this new kind of warfare without leadership at the very highest levels and, most important, from the president.

"I simply do not understand why our government is as engaged in matters involving objectionable behavior by North Korea, Iran, and even China, all of which I think are legitimate issues, and silent on Russia," said Carter. "I can't explain that."

The advent of the Shadow War should have surprised no one. In military terms, hybrid warfare is a natural product of a world with a single superpower and other rising or declining powers eager to challenge that superpower. For China, Russia, and other US and Western adversaries, hybrid warfare is the only way to take on a country such as the United States with otherwise unchallenged military might. In other words, the so-called gray zone is the only field of conflict on which these adversaries believe they stand a chance of winning.

John Scarlett was the head of Britain's foreign intelligence service, the MI6, from 2004 to 2009 and, before that, the head of its Moscow station. He explains the drive—in fact, the need—for hybrid warfare from Russia's perspective.

"It's not too difficult to understand what's been happening," Scarlett told me. "In very short order, we saw the humiliation, the resentment, the feeling that things are happening without Russia's interests being taken into account, the super-awareness of the power difference between the US and Russia.

"If you want to be treated as equal, you have to find some other way of expressing that," he said. "[Hybrid warfare] enabled much weaker countries to effectively take on much stronger countries. There's a natural asymmetry."

Despite employing similar strategies, Russia and China are different kinds of adversaries. China is a rising power, with growing territorial, economic, and military ambitions destined to conflict with those of the world's most powerful nation. Beijing sees itself in a long war with the United States for global dominance.

Russia is a declining power. With an economy smaller in purely GDP terms than some US states, the Kremlin knows it will never truly compete with the United States for global leadership. It sees the competition more as a zero-sum game: America's loss is Russia's win, and vice versa.

"With the Soviet Union, we're talking about collapse," said Scarlett. "With China, we're talking about rapid change and development and progress, as well as a high awareness of the fragile nature of the progress.

"It was always predictable in that growth that of course [China] would turn away from self-absorption into—on the international level—self-confidence," Scarlett continued. "We see a degree of assertion and assertiveness coming in, beginning regionally and becoming more widely internationally."

Russia and China, Scarlett believes, end up clashing with the United States on similar terms, though they arrived from different directions and are proceeding on different trajectories.

Russia's and China's Shadow War on the United States is driven by the same crucial and immutable forces, however, and those similarities could potentially lead to disastrous outcomes.

The first similarity is strategic: challenging US dominance in Europe and Asia serves Russian and Chinese ambitions to wield greater influence in their own regions. Each envies America's Monroe Doctrine—that is, the exercise of absolute power within America's close environs—and are working to establish their own versions of it.

The second common force is political. Both Moscow and Beijing suffer from a crisis of legitimacy at home. Their leaders are not elected by their people and therefore have little claim to power other than the fact that they hold it. In the modern age, no amount of censorship and propaganda can keep Russian and Chinese citizens from observing that Americans do choose their own leaders. Therefore, their best defense against their own populations is to portray the US political system as broken and corrupt—at least, no less so than the Chinese and Russian political systems.

The third force is arguably the most powerful. In undermining the United States, both China and Russia are seeking to right the wrongs of history and restore what they perceive as their countries' rightful positions as world powers. For Russia, the sin is recent: the collapse of the Soviet Union, followed by what Russians see as its subjugation by Europe and the United States. For China, the sin goes back generations, beginning with China's humbling in a series of nineteenth-century wars that over time, in their view, led its territory and economy to be similarly subjugated by the West.

In short, the Shadow War has all the ingredients of a real shooting war.

Russian and Chinese leaders are extremely aware of each other's history. Studying the collapse of the Soviet Union is mandatory among Chinese Communist Party leadership. And Mikhail Gorbachev is as reviled a figure in Beijing as in Moscow. Unlike in the United States, where he is seen as a Russian leader who helped avert World War III, in Russia and China he is seen as a leader who allowed his country to fall apart—and the United States and the West to pick up the pieces.

Today military planners in Moscow and Beijing openly discuss a whole range of unconventional means to reduce America's military advantage and influence in peacetime and, if need be, in war. For them hybrid warfare is not just asymmetric, it is never-ending. As General Gerasimov wrote, ominously, Russia's strategy involves creating "a permanent front through the entire territory of the enemy state."

Russia's interference in the 2016 presidential election extended the "permanent front" inside the US political process. Today US intelligence and congressional investigators are discovering that interference was far more extensive than originally thought. In addition to hacking and releasing the emails and other communications of Democratic Party and Clinton campaign officials, Russia engineered vast troll farms to influence millions of voters in swing districts with fake news and divisive stories. With the approach of new elections in 2018 and 2020, Russia took more alarming steps, threatening the very voting process itself.

US defense and intelligence officials now speak openly of the danger of repeating the errors of the 1930s, that is, observing aggression by adversaries in Europe and Asia while assigning false limits to those adversaries' ambitions. Those fears of repeating the mistakes of history are fueling calls to defend against the Shadow War now or face the danger of a wider conflict in the years to come. And yet, without a commitment throughout all levels of the US government, the United States faces the alarming prospect of emerging from the Shadow War diminished and defeated.

CHAPTER 2

Opening Salvo

(RUSSIA)

Citizens of Estonia, the tiny Baltic nation perched precariously on the border with Russia, poke fun at themselves for being the boring men and women of northern Europe.

"Our neighbors have numerous jokes about us being slow and not emotional enough for them," an Estonian journalist, Jaanus Lillenberg, told me.

So, in late April 2007, when the Estonian capital of Tallinn was rocked by violent street protests, the chaos was a shock. During those cold, rainy days in April, Tallinn was engulfed in violence.

"They broke windows, attacked cars which were parked at the roadside, threw stones, bottles, everything like that," said Lillenberg, who was then working in technology for the Estonian morning daily newspaper *Postimees*. "It's absolutely something that has never, ever happened in our history."

To residents of Tallinn, the scenes in the streets appeared almost otherworldly. Estonia is a low-crime country. The streets are clean. Estonians are passionate, but more about technology than protests. Public demonstrations are staid affairs. The chaos seemed borrowed from a dark fairy tale, or—more likely—from the streets of its "more emotional" allies and neighbors.

"We've seen it on TV—in Paris or Stockholm, or the United States—but never in Estonia," said Lillenberg. "This was almost like a story from the North Pole."

Riot police struggled to restore order but could not bring the crowds under control. Nearly two hundred people were injured. More than one thousand were arrested—a remarkable number in a city of fewer than half a million people.

Estonia has a long history, but it is still a young nation, having regained its independence from the collapsing Soviet Union only a generation ago, in 1991. Its separation from Russia, along with its Baltic partners Latvia and Lithuania, angered Moscow, stirring bitter memories of an empire lost. The wound still festers in the Kremlin. As the street battles unfolded, witnesses took notice that the rioters had one name—one country—on their lips.

"The crowd was shouting: 'Rossiya, Rossiya,' which means Russia, of course," said Lillenberg.

The spark for the protests had been the Estonian government's decision to move a decades-old Soviet war memorial. The Bronze Soldier, as it was known, honored Soviet Red Army troops who died fighting Nazi forces in Estonia during World War II. The memorial had become a gathering point for Russian and Estonian nationalists in sometimes violent demonstrations in the preceding weeks and months. For Russians it represented victory over the Nazis and Russia's proud past. For Estonians it was a painful reminder of decades of repression after their country's absorption into the Union of Soviet Socialist Republics, or USSR.

After moving the statue from Tallinn, the government planned to place it at a military cemetery outside the city center and exhume the bodies of unidentified Soviet soldiers buried nearby to give them a proper burial. However, many ethnic Russians in Estonia, and their compatriots across the border, interpreted the move as an insult to their Russian heritage and, more consequentially, as a further demonstration of Estonia's independence from Russian influence. To inflame tensions still more, the Russian media was circulating fake stories on social media and news websites that Estonia planned to destroy the memorial altogether.[1]

Estonia's foreign minister Sven Mikser, then a young member of parliament, recalled a growing sense of fear among his countrymen.

"The scale, a dozen or so cars being turned upside down. I mean, people, well, get nervous when things like that happen," said Mikser.

Few Estonians believed the riots were spontaneous. They suspected they were orchestrated by the Russian government.

"Absolutely," said Lillenberg. "It's like, no, it just doesn't happen."

As the hours passed and confusion grew, the violence on the streets would turn out to be just one front of a broader attack. Silently, in cyberspace, an invisible army was an offensive that would foreshadow later cyberattacks on Western Europe and the United States. The first clues were confusing and difficult to piece together. Lillenberg and his team watched the cyber assault unfold in waves and waves of seemingly innocuous online interactions. The target? His newspaper's comments section, which was being inundated.

"We had anonymous commentary and we usually got like eight to nine thousand comments a day. But then suddenly, it was like more than ten thousand in ten minutes," he said. "I was like, 'What?'"

Beyond the unprecedented pace, Lillenberg noticed an odd uniformity to the comments coming in. He and his colleagues identified and counted a handful of identical messages that were being repeated over and over again, by the thousands. It was becoming clear this was the work of computer bots rather than actual readers.

"There was a variation of about thirty distinct messages," he said. "It was the repetition. And it looked like there were no people behind it because people can't enter comments so quickly.

"That was the first thing, when we understood that something is wrong," he said.

Both the pace and scale of the attack would quickly accelerate. Within an hour, the number of comments flooding his newspaper's website jumped again, by a factor of ten, to 100,000 in ten minutes.

The assault on the *Postimees* website was being repeated across the private and public sectors. Jaak Aaviksoo, just two weeks into his job as minister of defense, immediately took note. Like everyone in Estonia, he was educated in the power of technology.

"I looked at different news portals and they were down. I asked, what's going on? And it was reported the banks were down, government websites were down," said Aaviksoo.

Seated in an office from which he had yet to remove the furniture or artwork of his predecessor, Aaviksoo suspected a coordinated attack from abroad.

"That was clear that it's not bad weather," he said. "It's bad people out there."

Estonians, perhaps the most connected population in the world, suddenly found themselves cut off—with no access to news or government websites and thus no information about what was going on. Electronic banking, which at the time accounted for the vast majority of all financial transactions in Estonia, was also down. The attack exploited a glaring vulnerability for the country. Tiny Estonia, known for the medieval walls and cobblestoned streets of its capital's old city, is a technology powerhouse: the first country to allow online voting and the birthplace of Skype. But now one of the most wired countries in the world was under one of the most crippling cyberattacks the world had ever seen.

Estonia had become the first victim of a state-sponsored cyber-attack on another nation. The attack took the form of a "distributed denial of service," or DDoS, attack. DDoS attacks were not new at the time, but the size of this onslaught was unprecedented. Russian hackers had hijacked tens of thousands of computers in more than one hundred countries and unleashed them, unbeknownst to their owners, on targets across Estonia.

"Think about a huge shopping mall," explained Lillenberg. "People go in, buy some stuff, and go out. The same thing goes for the Web servers. A user comes in, asks for stuff, the server gives some stuff, and the user gets out. That's how the flow goes.

"Imagine your shopping mall can handle like ten thousand or fifteen thousand people. But now imagine two million guys pressing on the front door with absolutely no intention to buy anything, just to block the front door. That is a DDoS attack," said Lillenberg.

Precisely because Estonia was so far advanced, it was extremely vulnerable.

"We learned that the cyberspace constitutes part—an integral part—of the critical infrastructure," said Mikser. "So we have to defend and keep these systems going even in times of crisis, even in times of attacks."

The attack, however, was playing out on multiple fronts: riots on the streets, botnets on the Web. It was hybrid warfare in action. Together these secret armies appeared tasked to paralyze the country. As a student of Soviet military tactics, Defense Minister Aaviksoo saw the hand of his country's neighbor Russia.

"Street riots to an extent we had never seen before. Coordinated and concentrated attacks in cyberspace. This was a wake-up call for Estonia," Aaviksoo told me. "Of course, we found that these attacks were not that spontaneous as they pretended to be. Coordinated, focused, global. It clearly indicated that there had been considerable resources involved to make these attacks possible."

The cyberattack was the largest ever launched by one state on another. The addition of orchestrated riots on the ground added an alarming element of physical violence. That Estonia was a member of NATO made this an overt challenge not just to one country, but to the United States and Europe as well. The attack on Estonia in 2007 was the opening salvo in the Shadow War.

Some Estonians began to fear that the riots and cyberattack were laying the groundwork for a full-scale invasion. Estonians were well aware of their neighbor's deep dissatisfaction with losing its client states in Eastern Europe. The Baltics, including Estonia, which shares a border with Russia, were a particularly sore point, having been the first to be absorbed into the Soviet Union and the first to declare independence following the USSR's collapse.

"It is pretty much the purpose of a very strong DDoS, a full-scale assault just to put a country down for a while, information-wise," recounts Lillenberg.

Like their counterparts in Ukraine, Georgia, and other former Soviet republics, Estonians had long been the target of Russian propaganda. Supporters of independence were dismissed as nationalists and fascists. Ethnic Russians were portrayed as victims in need of Russian help. Nearly sixteen years since Estonia had regained its independence, memories of Soviet domination remained fresh and the wounds raw.

"There was something I saw. Everyone was trying to be strong, but everyone is also a person, a human being. There was some level of worry that was very personal," said Lillenberg. "They were worried about themselves and their families. We were very, very worried.

"If you are familiar with Russian doctrine, it's a matter of escalation. At some points there are tanks, at some points there are nukes, but it's all one long line and it starts with creating and distributing fake news and then it escalates," he added.

Such fears were not confined to panicked members of the public. Defense Minister Aaviksoo made urgent calls to his combat commands. They reported no incursions into Estonian airspace or onto Estonian territory, including on its well-defended eastern border with Russia. Still, the Ministry of Defense put itself on a war footing, against an enemy that had yet to be identified.

"What was important was that, at least psychologically, it posed a national security threat," Aaviksoo told me. "A large proportion of population was afraid, destabilized. No human casualties, no material loss, but the clear understanding was that we are under attack.

"The real fight is going on in the psychological space—between the ears of people, in their minds," he said.

This was an attack on the psychology of a nation and its people: confuse, divide, antagonize, frighten, and sow doubts about their leaders.

"People started asking questions as to whether the government is in charge," he added. "What's going on?"

As the government struggled to calm the public, restore order, and repel the cyberattack, it was confronted with a series of critical questions as yet without answers. A faceless enemy was systematically shutting down their country. Estonia was suffering the equivalent of a cyber blockade, cutting Estonians off from virtually every public and private service and, soon, cutting their country off from the outside world. Members of the public and the government had only one suspect in Russia. However, neither the rioters in the streets nor the bots on the Web wore uniforms. Was Estonia at war? And if so, with whom?

To Defense Minister Aaviksoo, a war did not require invading troops and falling missiles.

"A common understanding is that it depends not on the means you carry out an attack, but on the impact. If there is extensive material damage, the loss of lives, injuries," explained Aaviksoo. "So, if the impact is of that scale, it's an act of war.

"Whether it was a missile or a cyberattack, it doesn't matter," he added.

Russia, with all its first-world military capabilities, including a nuclear arsenal more sizable than that of the United States, was borrowing tactics from rogue states and nonstate actors. It was attacking tiny Estonia via asymmetric means. Estonian officials compare 2007 to the September 11 attacks.

"With 9/11, many people called it a low-tech, high-execution attack," says Mikser. "Similar with a cyberattack like this, when you drop your guard, when you're not careful, then obviously, the attackers will abuse it. They will take advantage of that."

Estonian leaders considered these questions carefully at the same time they were attempting to defend their country. But these were not questions for Estonia alone. Estonia belongs to NATO, whose members are bound by treaty to view an armed attack on one nation as an armed attack on all and to mobilize to come to their allies' defense.

Article 5 of the NATO Treaty states: "The Parties agree that an armed attack against one or more of them in Europe or North

America shall be considered an attack against them all and consequently they agree that, if such an armed attack occurs, each of them . . . will assist the Party or Parties so attacked by taking forthwith, individually and in concert with the other Parties, such action as it deems necessary, including the use of armed force, to restore and maintain the security of the North Atlantic area."[2]

Estonia was being forced to define and interpret the laws of modern warfare in real time. Did a cyberattack coupled with orchestrated protests in the streets trigger a NATO response if there was significant loss of life and damage to property? Some believed nonmilitary action only qualified if it triggered a loss of life equivalent to military action. This was a question that might require a new or altered definition for new threats in a new era of warfare.

In the end, Estonia did not ask its NATO allies to respond militarily. Aaviksoo says they simply kept their allies informed.

"Our response was that we told everyone that we are under attack. We shared our experience. We informed our friends and neighbors," Aaviksoo said.

For its own part, Estonia chose not to retaliate, either. The Estonian government and defense establishment were focused on repelling the cyberattack, achieving calm in the streets, and getting their country back online.

"We never retaliated in the direct sense of the word," he said. "But I think in every conflict there must be the capability to contain a possible opponent. You have to be able in a credible way to demonstrate your ability to hit back. That must be there. You cannot give up that credible ability to hit back."

Five days into the attack, Estonia took one bold step: publicly naming and shaming its powerful neighbor Russia as the culprit. The Estonian foreign minister at the time, Urmas Paet, said his country had electronic evidence leading all the way to the Kremlin.

"It has been established that cyber terrorist attacks against Estonian governmental institution websites and that of the President's Office have been made from IP addresses of concrete computers and by concrete individuals from Russian government organs, including the administration of the President of the Russian Federation," Foreign Minister Paet said in an official statement released May 1, 2007.[3]

Foreign Minister Paet declared the cyberattack an assault, and not just on Estonia, but on all of Europe.

"We consider the European Union under attack by Russia because Russia is attacking Estonia," Paet continued. "The attacks are psychological, virtual, and real."

Psychological, virtual, and real. Paet's words were a powerful description of the Shadow War in action. Estonia's experience haunted and mobilized Estonia's military planners to defend Estonia against similar attacks in the future, and to warn their NATO allies, including the United States, of what was likely to come their way.

Estonia's focus remained on defense and recovery. At the *Postimees* newspaper, Jaanus Lillenberg and his colleagues launched a small but nimble cyber rearguard. With email networks down and before the wide use of Twitter and Facebook messaging, they coordinated their strategy via text messages, beginning with simple cyber tools.

"First of all, we put a limit on how many comments you could enter from one IP address," he recalled. "Then we built this very fast, very smart filtering system in just a few hours."

The system they developed would filter out comments that included certain key words and phrases that Lillenberg noticed in the flood of botnet-generated comments, such as "fascists" or the "SS." The bots, like today's Russian Twitter trolls, were stoking conspiracy theories, including fake stories that the Estonian government planned to destroy the Soviet war memorial. He and his team were writing the code for these new filter programs in real time, while their computer systems were under attack. It was a seat-of-the-pants, all-hands-on-deck operation.

"I had a guy," remembers Lillenberg, "a software developer who had a flu called me at three o'clock in the morning to say, 'Yes, I think I got it working.'"

Their system slashed online traffic enormously.

"That last system really, like, killed it," Lillenberg told me with a smile.

Lillenberg and his team developed another tool designed to confuse the bots. Judging—correctly it would turn out—that the attack was developed by Russian speakers, rather than native speakers of Estonian, Lillenberg had an idea: giving visitors to the site a simple test.

"We didn't want to use one that was already public, because they would have the means to go around it," he explained. "So we wrote another one which was very, very stupid, but at least it wasn't something you could find on the internet.

"You have three icons," he said. "Let's say, like, scissors, a watch, and a plane. And there is a question in Estonian that asks, 'Please click on the plane.' And so, if the visitor—or the bot, really—doesn't understand Estonian, he's wondering what to do with those icons."

It wasn't quite Bletchley Park breaking the Enigma code. However, their simple fix worked, presaging the kind of fixes cyber-security experts would deploy to defeat similar DDoS attacks in the months and years that followed.

"It took some time. I would say forty-eight to fifty hours of work for us to get it under control," he said. "We sorted things out, set the guys to work, and expected some results."

Lillenberg's fix was one small victory in one small skirmish in a long, drawn-out cyber onslaught. The low-tech, high-impact assault on Estonia would continue for weeks. Eventually Estonian leaders were left with one punishing option to stop the onslaught: blocking all international Web traffic, in effect temporarily disconnecting one of the most connected countries from the rest of the world.

"No one from the outside world could get any information from Estonia. I think that was also the aim of the attack, actually," said Jaanus Lillenberg. "If you have a closed area where the information doesn't go in or out, you can do a lot stuff there. Military operations, information operations."

————————

Looking back, Russia's 2007 cyberattack on Estonia incorporated elements that would characterize similar attacks in the years that followed on both former Soviet client states, such as Georgia, and—later—on Western nations, including the United States.

First, Russia employed cyber weapons that were expansive but relatively blunt, in this case a DDoS attack designed to overwhelm networks and shut them down. The size was unprecedented: hijacking tens of thousands of computers in more than one hundred countries. However, the tools were far from sophisticated.

Moreover, while Russia had not declared war, its hand in the attack was easy to identify. For one, the cyber portion of the assault coincided with physical action on the ground, in this case pro-Russian protests, which Estonian officials believed were coordinated with the help of the Russian authorities. In addition, despite the fact that the botnets operated from dozens of countries, they were marked by electronic fingerprints, including ties to Russian IP addresses and code written in the Russian language.

More broadly, Russia was revealing an essential part of its grand plan in the Shadow War: undermining the West by sapping confidence in the Western system as a whole.

"Russia has its own strategic interest, which is—and which they define as—diametrically opposed to the strategic vision of the Western alliance," said Mikser. "So they have been using different ways to divide the West, create confusion, and basically undermine the confidence of the societies, of people, in the democratic processes."

Today, more than a decade later, 2007 remains a defining moment for Estonia and its leaders. Just as the September 11 attacks transformed the US intelligence community's approach to terrorism, Russia's unprecedented cyber assault generated a broad rethinking of Estonia's cyber vulnerabilities and how to minimize them.

"It was first time ever in history anybody was undertaking [such an attack]," Estonian president Kersti Kaljulaid told me. "And it was possible, because Estonia was a digital state. You couldn't attack any other state this way, at that point. So it was a historic moment, of course."

At just forty-six, Kaljulaid is the youngest-ever president of Estonia. In person and in public, the mother of four exudes Estonia's no-nonsense attitude in standing up to its much larger and increasingly aggressive neighbor. Like many Estonian officials and citizens I have met, she expresses no fear, just a sense of purpose and conviction. That sense of purpose is apparent in the numerous measures and investments Estonia has made since 2007 to defend itself. Estonia has become a sort of "cyber Beirut"—perpetually surrounded by the Shadow War and under threat of being engulfed by it, and yet somehow managing to survive and even thrive.

Denial-of-service attacks, like the one that paralyzed the country in 2007, are now commonplace, but Estonia's defenses have rendered them largely ineffective.

DDoS attacks, President Kaljulaid told me confidently, have become "like rain." "Nobody notices these drops falling on our systems," she added.

Of course, ten years is a lifetime in technology terms. Russia, China, North Korea, and other cyber actors have adjusted and advanced their cyber capabilities. And with those advanced capabilities have come more aggressive cyberattacks.

"It shows you how the technology has developed and how defenses have developed, but also how much more active these aggressive actions [have become] on the internet sphere," she said.

And yet, as the sophistication of cyberattacks has advanced, so has Estonia's ability to defend against them. Its record stands for itself. Estonia is unusual in that it did not suffer major losses from two of the most damaging cyberattacks of the last several years: the 2017 "WannaCry" ransomware attack, which the United States attributed to North Korea, and a global attack on network infrastructure in 2018, which the United States blamed on Russia.

"They did not actually achieve damage in Estonia simply because our people are more cyber hygienic than elsewhere. They keep themselves safe in digital sphere," said President Kaljulaid. "In fact, we are seeing less, I believe, of the obstructive cyber activity because Estonian society is at the higher level of cyber hygiene and probably we are more difficult to attack."

Remarkably, Kaljulaid says that Russia has found Estonian cyber defenses so impenetrable that it is no longer even trying.

"We have been preparing ourselves . . . to avoid these attacks, and guess what, we haven't seen a single one," she said.

Its success, Estonian leaders emphasize, would not be possible without a nationwide awareness of the threat and a nationwide effort to defend against it.

"You have a national responsibility to explain to people that they need to take their individual responsibility," President Kalju-laid emphasized. "Indeed, technology will never protect people."

Every Estonian official you meet champions the wisdom and necessity of "cyber hygiene." One irony of even the most dam-aging cyberattacks of the last decade, including Russia's interfer-ence in the 2016 US election, is that they used relatively blunt tools requiring simple user error. Phishing attacks, like the one that successfully targeted the Hillary Clinton campaign, required her campaign chairman to actively click on a link. Estonians are trained—browbeaten, even—never to make the same mistake.

"Cyber hygiene, cyber hygiene, and cyber hygiene," repeated President Kaljulaid. "We teach our people. It's essential."

Their cyber education begins in schools with so-called Web po-lice, who educate children to avoid unknown actors encountered on the Web with the same caution they would a stranger in the playground. As Kaljulaid said, technology cannot protect people from cyber threats. People must learn to protect themselves.

Estonia's success is remarkable considering how much Estonians do online, from receiving public support payments to banking to voting. Even with Russia's subsequent interference in elections across Europe and in the United States, Estonia has never wavered from its online voting system. The stakes couldn't be higher. A cy-ber breach could undermine confidence in Estonia's elections and financial system. As a result, digital signatures are the norm for all kinds of online transactions.

"Our people know that if something is not digitally signed, it's not safe," said Kaljulaid. "If somebody has signed digitally with the digital identity and sent you the information, you can be sure that it is safe, and you can also be sure that it was encrypted when it was signed. So safe internet exists. Everything else, our people know, is unsafe."

The Estonian government is not, however, placing all its faith in the vigilance of average Estonians. Estonia is taking aggressive steps not just to prevent the next attack, but to ensure that any attack cannot paralyze the country as Russia managed in 2007. One such measure is the establishment of so-called data embassies.

A data embassy is a heavily protected collection of servers located outside of Estonia, containing a giant digital copy of all Estonian government data, from government communications to voter data to financial and health records. The idea is to back up the entire Estonian government outside the country, so that Estonia can access the data in the event of a disabling cyberattack at home. When announcing its plans in June 2017, the Estonian government said, hopefully, "Estonia's pilot project could, again, set an example to the rest of the world."[4] Estonia opened its first data embassy in Luxembourg in 2018.

"It enjoys all the rights according to the bilateral agreement between Estonia and the other country of an embassy," President Kaljulaid explained. "So it's, as all other embassies, technically our territory; only we can enter it and give permission to enter it."

Estonian private sector companies are reaching out for foreign help as well.

News outlets, for instance, have engaged foreign partners to stand up mirror images of Estonian news websites on servers outside the country. Estonian Public Radio is one of them, though for security, it will not identify in which countries those backups are based.

"I cannot say the places, but we have a few good media houses which now have geographical mirrors of the news site," said Lillenberg, who now works for Estonian Public Radio.

Lillenberg, now a grizzled veteran of the 2007 cyberattack, describes these digital outposts in military terms.

"So, if they fired their new cruise missile and hit the news building, nothing happens because the other building in the other geographical location is up and running all the time," he said.

In the ongoing cyber conflict with Russia, Estonia's private sector is very much on the front lines. In fact, the Estonian government relies on private companies as citizen soldiers.

"The other thing we learned was that actually much of the expertise that's necessary for dealing with these threats lies in the private sector," said former defense minister Aaviksoo. "And so it's important to cooperate as closely as you can between the government and the private sector."

To emphasize that partnership and to highlight the essential role of Estonian citizens in cyber defense, Estonia has stood up Estonian Defence League's "Cyber Unit." Composed of volunteers with essential information technology and cybersecurity skills as well as expertise in law, economics, and more, they train regularly to help fight off a crippling cyberattack in the event of a crisis. The unit has also allowed the government to draw on private sector talent it could never attract full-time.[5]

The parent Estonian Defence League is a volunteer militia that was started in 1918 and reestablished with Estonia's independence from the collapsing Soviet Union in 1991. Estonians say the league was inspired by the Minutemen of the American Revolutionary War. And today, armed militia units of the league train regularly to help fend off a conventional attack. Members of the Cyber Unit are the "techie" Minutemen of the cyber battlefield, operating as a reserve unit of private sector volunteers to be called up to defend the nation when under cyberattack. Estonia's NATO allies are studying the Cyber Unit as a model for their own countries.[6]

Despite their success in surviving 2007, Estonian leaders emphasize that the Shadow War extends far beyond cyberspace. And it is Russia's attacks via other means that are more difficult to defend against, especially information operations such as attacks on Western elections.

"This is much more dangerous I believe than cyber conventional, because cyber conventional can be solved by technical systems and good cyber hygiene," President Kaljulaid told me. "We have to strive—we have to explain to our people that this is the case."

"Hybrid scenarios are something that are fashionable now following the annexation of Crimea and the aggression against Eastern Ukraine," said Mikser. "Actually, we have been experiencing this sort of hybrid pressure ever since we regained our independence.

"There's always been a combination of political pressure, psychological warfare, if you wish, and also economic measures being applied, attempts to interfere in our domestic political affairs," he said.

In the age of the Shadow War, Kaljulaid says the West must look at Russia's attacks as a whole, rather than individually—and fight them as a whole.

"All Russian actions, cyber or physical, starting from occupying part of Georgia, then moving on to Ukraine, then testing our cyber defenses, attacking democracies," she said. "These are all elements and signs of trying to turn the tables on our rules-based [system]. We need to see it in a holistic way, as a complete process."

That requires unity among Western allies and a willingness to call out and punish Russian behavior. As a powerful example, Kaljulaid cited the moment when French president Emmanuel Macron confronted Russian president Vladimir Putin on the steps of the Élysée Palace in Paris regarding Russian interference in France's presidential election.

"You see President Macron beside President Putin saying, 'You did this to our democratic election,'" she recalled. "These are things we people see. They are the things that are big enough to catch attention."

Even with such warnings, Russia has carried out even more aggressive attacks on the West, opening new fronts in the Shadow War. Many European diplomats and officials reacted with particular alarm to Russia's poisoning of a former Russian spy Sergei Skripal and his daughter, Yulia, on the streets of Salisbury, England, in March 2018. The weapon, which police believed was smeared on the doorknob of the Skripals' flat, was the Russian-made nerve agent Novichok. Three months later, a British couple, Dawn Sturgess and Charlie Rowley, who were unrelated to the Skripals, was poisoned with the same nerve agent, police believe, after handling a contaminated container. Sturgess would later die.

"This is a physical action on the territory of a NATO country, unprecedented in NATO's history," Kaljulaid told me. "My question is, what might be next? We need always to think, what might be next? And be ready."

LESSONS

Russia's 2007 cyber assault on Estonia provided two sobering lessons for the United States and the West. First, the assault showed that even a relatively blunt cyber weapon can paralyze an entire nation. The "distributed denial of service," or DDoS, attack Russia unleashed on Estonia required minimal cost or complication and is a tactic easily replicated and deployed by a range of smaller state and nonstate actors. Second, its 2007 attack demonstrated that Russia was willing to launch cyber weapons against Western nations with the intent to disrupt and undermine on a scale and to a degree it had not done so before. Coupled with an elaborate influence operation on the ground in the form of orchestrated protests and the spreading of false information, Russia created panic and division inside a foreign adversary, an early hint of the campaign of disruption it would later launch against Western Europe and the United States.

The United States and its Western allies largely missed those lessons in 2007 and therefore missed signals of what was to come in the decade that followed. This pattern of missed warning signs would continue, even as those warnings became clearer and more threatening. Throughout, US and Western leaders and policy makers persisted in their mistaken view that Russian leaders wanted, to a large degree, what the West wanted: a mostly friendly relationship governed by the international rules-based order created and defined by the West. This included the adherence to treaties designed to minimize the danger of military confrontation and allow for some military cooperation. This mistaken expectation became part of the conventional wisdom regarding Russia. The events that followed proved it to be wishful thinking.

CHAPTER 3

Stealing Secrets

(CHINA)

To his American friends and contacts, Stephen Su was an affable businessman and gregarious guy.

"People liked him," Bob Anderson, the FBI's former head of counterintelligence, told me. "They didn't think he was an asshole and I know that sounds stupid, but people are people and that's how it started."

Stephen Su, who also went by his Chinese name Su Bin, lived in his native China but traveled frequently to the United States and Canada, to build a business in the aviation and aerospace sectors. His company, Lode-Tech, was a small player in a field of giants. It focused on manufacturing aircraft cable harnesses, a product safely at the low-tech end of the military aircraft sector. However, over the course of some five years, from 2009 to 2014, Su steadily and deliberately built a network of close business contacts inside far bigger US and Canadian defense contractors with some of the most sensitive US military contracts. As Anderson explained, Su made it a point to get to know the people who had access to those technologies, or people who knew the people who had such access, and "getting them to trust him."

His contacts described Su as an ideal partner, intent on making deals that benefited not only himself but also his American and Canadian counterparts. He was all about business, but he was also good company. Over the years, he enjoyed dozens of expensive dinners over wine at some of the best restaurants in Seattle, Vancouver, and Los Angeles.

"So, he cultivates you so over time," Anderson recalled. "First, you're just buddies and then after that, 'What are you working on? What are you doing? Boy, that's really interesting.' Then, in a lot of these cases he'll talk about, 'Well, you know, there's a great way that we can make money at this.' Or 'There's a great way that me and you could become partners with different individuals I know that would want access to information like this.'"

The information Su was most interested in related to three of the most advanced US military aircraft ever built, the Lockheed Martin F-35 and F-22 stealth fighters and the Boeing C-17 Globemaster transport aircraft. Though they were the products of two of the Pentagon's biggest military contractors, each drew on thousands of components sourced from dozens of smaller suppliers. That supply chain provided numerous ins for Su—as well as a convenient explanation for any partners who grew concerned about the kind of information he was looking for.

"Su would say, 'I'm not asking you to give me the F-35, but what's it matter if I get one system out of it that we could sell to a friend or a perspective client?'" said Anderson. "And then go from there, and it takes time."

Unbeknownst to his contacts, Su wasn't working alone. In fact, he was part of a three-person, cross-border team, with Su in North America, and his two partners—identified in the FBI's 2014 criminal complaint only as "uncharged co-conspirator 1" and "uncharged co-conspirator 2"—from mainland China. According to the FBI, Su would identify valuable computer files inside the target companies, then transmit that information to partners in China, who would hack their way into the target companies' computer systems to steal the identified files. The team would then sell the stolen files to interested parties in China, namely state-owned enterprises in the military sector. As the criminal complaint noted, they did so not only at the behest of the Chinese government but also "for their personal profit." This was espionage for both country and their bank accounts.[1]

Emails later obtained by the FBI showed that their modus operandi was simple and efficient. The team came together for the first time in the summer of 2009, when Su sent the first emails to his co-conspirators identifying potential targets inside the United States. In an email dated August 6, 2009, Su attached a password-protected Excel spreadsheet containing the email addresses, telephone numbers, and positions of some eighty engineers and other personnel working on a new military project. Su's tradecraft could be low-tech, even clumsy. The subject line for the August 6 email was "My Cell Phone Number," which the FBI later discovered indicated that the password for the protected Excel file was simply the number for Su's cell phone.[2]

Four months later, on December 14, 2009, Su sent a similar email, this time with the subject line "Target," listing the names and positions of four other executives, including the president and vice presidents of a company that manufactured weapons control and electronic warfare systems for the US military. Later FBI analysis would determine that the divisions identified in those early emails matched targets later hacked by Su's team.[3]

The hackers' next step resembled methods used by Russian hackers to penetrate the Democratic Party during the 2016 US presidential election. The hackers sent so-called phishing emails to employees of the target company designed, as the FBI explained, "to appear as if it came from a colleague or legitimate business contact." If the recipient clicked on a link contained in the email, or opened a document attachment, an "outbound connection" would be established between the victim's computer and another in China under the hackers' control. The hackers would then install malware on the victim's computer, allowing them to control the computer remotely and—more alarmingly—explore the company's entire network.[4]

Su and his team took careful steps to conceal the origin of their cyber intrusion. To do so, the hackers' outbound connection from the target company would be routed through a series of servers in a number of different countries around the world. These "hop points," as they are known, would obscure who was doing the hacking and where they were operating from—if and when the hackers were discovered.

As they wrote in a 2013 internal report obtained by the FBI, "In order to avoid diplomatic and legal complications, surveillance work and intelligence collection are done outside China. The collected intelligence will be sent first by an intelligence officer via a preordered temporary server placed outside China or via a jump server which is placed in a third country before it finally gets to the surrounding regions/areas or a work station located in Hong Kong or Macao."[5]

The final step in their theft—that is, the final "hop" back to their clients in mainland China—did not go over computer networks at all. Su and his partners set up what they referred to as "machine rooms" in Hong Kong and Macao where the stolen intelligence would be collected and then carried across the border into China by hand.

"The intelligence is always picked up and transferred to China in person," they wrote in a 2013 email.[6]

As it turns out, Su and his partners would have unfettered access inside Boeing's network for three years before the intrusion was first discovered. During that time, they would claim to have stolen some 630,000 digital files—totaling a gargantuan 65 gigabytes of data—on the C-17 alone. They stole tens of thousands more files on the F-22 and F-35.[7]

Su Bin's team, while enormously successful, was just one small part of a massive army of Chinese hackers dedicated to stealing America's most sensitive government and private sector secrets. Over the last two decades, China has built an enormous infrastructure charged with cyber espionage. The Office of the US Trade Representative estimates that the United States loses up to $600 billion *per year* in intellectual property. Since it deems China "the world's principal IP infringer," the USTR believes China may be responsible for the bulk of those losses.

The theft of US secrets is one of the most insidious fronts of the Shadow War: constant, deeply damaging to national security, and happening in plain sight. During my time as chief of staff at the US embassy in Beijing, US firms—though aware of the theft—often refused to ask for government help, or even to identify cyber breaches, for fear of alienating their Chinese partners or losing access to the Chinese market altogether. In fact, China's strategy relies on—and cultivates—that fear.

One senior US law enforcement official described China's espionage apparatus to me as akin to a "tapeworm," feeding off tens of thousands of US institutions and individuals, to siphon away America's most treasured asset: its ingenuity. Beijing's goal is nothing short of surpassing the United States as the world's most powerful and most technologically advanced superpower. Chinese leaders would prefer to do so peacefully, but if there is a war, they want to level the battlefield.

This is not simply conjecture. It is reflected in the rhetoric of the highest levels of Chinese leadership. President Xi Jinping envisions China at the forefront of innovation by 2035 and, beyond this, as a leading global power by 2050. Noble goals to be realized, but ones that the leadership has shown it believes will necessitate some leapfrogging—and even cyber espionage—along the way.

"This is about world domination and when or if there has to be a conflict—and unfortunately there probably will be one—they want to be mano a mano, if not better than the US, and that's what they've set their sights on for the last thirty or forty years," Anderson explained.

Cyber espionage may seem like a softer, less bloody front of the Shadow War. However, Anderson says that Chinese security services operate as brutally in cyberspace as on any other battlefield.

"The Chinese are more vicious than the Russians," Anderson told me, pausing to make sure I was listening. "They will kill people at the drop of a hat. They will kill families at the drop of a hat. They will do it much more quietly inside of China or in one of their territories, but they absolutely—if they have to—will be a very vicious service."

B ob Anderson found his way into counterintelligence from corners of law enforcement where "vicious" comes with the territory. He started as a Delaware state trooper, policing crimes ranging from reckless drivers to drug dealers to homicide. He began his tenure at the FBI in 1995 on a drug squad in southeast Washington, D.C., when the nation's capital had one of the highest drug and violent crime rates in the country.

"We were buying coke, crack, meth, heroin," Anderson recalls. "D.C. was the murder capital of the United States then."

As he rose through the ranks, he served on a SWAT team and hostage rescue team, before being promoted to supervisor of the FBI's counterintelligence team in 2001. His experience on the streets was surprisingly relevant training for the world of international espionage. Foreign agents, Anderson remembers, were as violent and dangerous as the drug dealers he had chased on the streets of Washington. His Russian counterparts were particularly tough.

"Oh shit, they hate your guts," Anderson said. "They hate your guts 'cause you're an American and it's not 'cause you're white or black or male, female. They hate you 'cause you're an American."

Anderson told me the story of sitting down with a senior official of Russia's foreign intelligence service, or SVR, for a spy swap. It was the first time the FBI's assistant director of counterintelligence had ever sat down with his or her Russian counterpart on US soil.

"He brought two secretaries—obviously two giant GRU [Russia's foreign military intelligence agency] guys—that are stuffed in a suit, ready to break anybody's neck in the restaurant that we went to," he said. "It was like something out of a spy novel."

The seventy-two-year-old Russian official sitting across from him for the swap was, like President Putin, a veteran of the former KGB.

"This dude's probably killed more of his own people and he is just staring at me across the table and you can tell he doesn't give a shit that I'm an FBI agent," Anderson told me. "He just hates me 'cause I'm American."

The hate remains a constant, but from his post in counterintelligence, he watched how the very nature of spying changed, supercharged by the advent and expansion of cyber tools.

"I was arresting a shitload of spies and I started to see a change in how spies were operating," Anderson said. "Instead of being old dead drops under a bridge, everything was on a thumb drive. Everything was in the cloud and that's back when people didn't even know what the hell I was talking about."

Anderson tells his spy stories with a veteran cop's nonchalance as well as a touch of bravado. To him, a drug dealer from Washington, D.C., shares more than you'd imagine with a cyber thief from China. Both are willing to lie, cheat, fight, and even kill to get what they want. With his street credentials, Anderson found himself as the lead investigator on what would turn out to be one of the most devastating cyberattacks ever on the United States, including the massive breach of the NSA by Edward Snowden, North Korea's hack of Sony Pictures, and China's penetration of the US Office of Personnel Management, which would expose the personal information of millions of government employees who held, or once held, security clearances (including myself, I would later discover).

"We had six hundred people in 123 countries," said Anderson. "And when you start seeing all this crap, you start—at least I do—from a criminal background, seeing how, like cartels, they use money, but they use virtual currency now. And they can launder it through, I don't know, fifty countries in an hour. I mean how do you track that?"

The exact number of spies like Stephen Su is hard to pinpoint but Anderson estimates that, at any one time, there are dozens of teams like his operating in the United States. And behind them in China, Anderson says, are far more hackers at work, some employed full-time by Chinese security services, others working on a part-time basis. Call it a cyber "national service" program for young, well-educated Chinese.

"You'd go to jail here, but the Chinese have tens of thousands of young kids—like our MIT's or Stanford's best—hacking against the US," says Anderson. "They pay them to do that. That is quite routine for them.

"And they're very calculated in what they do, they have requirements just like the United States intelligence community has requirements," said Anderson.

They are also extremely ambitious in their goals. In a 2011 email, Su's team claimed with a flourish that the information they were stealing would "allow us to rapidly catch up with US levels"— "To stand easily on the giant's shoulders."[8]

In pursuit of their lofty goal, Su and his co-conspirators kept meticulous records—eager to prove their usefulness to their Chinese government clients and to improve their bottom line. And so, in a series of glowing self-assessments contained in emails, they provided an exhaustive account of their crimes. They were aggressive marketers of their stolen files. They were, after all, in the business not just of stealing sensitive information but selling it at the best price possible. To do so, they sent regular email updates on their work—updates that read like sales pitches, riddled with flowery self-praise.

On July 7, 2011, about one year after penetrating Boeing's secure network, Su Bin's "unindicted co-conspirator 1" emailed a report to his supervisor "unindicted co-conspirator 2" titled "Past Achievements," which included a long list of material stolen from US defense contractors. They claimed to have gained control of one contractor's servers and stolen twenty gigabytes of technological data. In addition to stealing files related to the C-17 and F-22 and F-35 aircraft, they bragged of conducting "reconnaissance" on files related to a US-made unmanned aerial vehicle, or UAV.

"We have collected a large amount of information and mailboxes of the targeted relevant personnel," read the July 7 email. "We have also obtained the password for the customer management system of the supplier and controlled the customer information of that company."

Other "Past Achievements" included non-US targets. "Through long-term reconnaissance and penetration, [we have] secured the authority to control the website of the . . . missile developed jointly by India and Russia," wrote one of Su's Chinese co-conspirators. He cited military technology stolen from Taiwan and political information stolen from the "Democracy Movement" and "Tibetan Independence Movement" inside China. They focused on any information they believed the Chinese government would want to know—and be willing to pay for.[9]

However, the United States was by far their prime target. On February 27, 2012, one of the Chinese co-conspirators sent an email with the subject line "Complete Listing." An attachment documented some thirty-two US military projects they had targeted along with the amount of technological data they claimed to have stolen on each project.

According to the FBI's 2014 criminal complaint, next to "F-22" was the figure "220M," for 220 megabytes of data. Figures next to some of the other thirty-one targeted projects were followed by a "G," which an FBI expert concluded referred to gigabytes of data.

A gigabyte equals one thousand megabytes, or the equivalent of about five thousand books.[10]

It was an extraordinary trove of information on some of America's most advanced and most sensitive military projects. Later FBI analysis would confirm that the file directories, technical schematics, and other proprietary information stolen matched "originals obtained [by the FBI] directly from US companies or US government entities."[11]

The years 2012 and 2013 would be extremely busy. Su and his co-conspirators claimed a series of successes as they repeatedly updated their Chinese clients on the progress of their theft.

On February 13, 2013, one of Su's partners again wrote his supervisor explaining, "The focus on the US is primarily on the military technologies but it also touches other areas whereas the focus on Taiwan is mainly on the military maneuvers and military construction.

"In recent years," he continued with a flourish, "we, with relentless work and through multiple channels, have obtains [sic] respectively a series of military industrial technology data including F-35, C-17 . . . as well as the Taiwanese military maneuvers, warfare operation plans, strategic targets, espionage activities and so forth."[12]

In August 2013, Su's team sent their most comprehensive report of their hacking activities in the United States. The glowing account includes some of their characteristic and colorful self-praise.

"We safely, smoothly accomplished the entrusted mission," they wrote. "Making important contributions to our national defense scientific research development and receiving unanimous favorable comments."[13]

The hackers documented the length and sophistication of their hacking. In one email, they reveal the date when they first penetrated

Boeing's internal network as January 2010. They say obtaining access to Boeing's "secret network" was difficult, as it was protected by "anti-invasion security equipment in huge quantities" with some eighteen separate network domains and ten thousand computers.

They took special precautions to avoid detection, including operating only during US work hours, adding that they "repeatedly skipped around in its internal network to make it harder to detect reconnaissance." They also note that they used "hop points" in at least three different countries, adding that "we ensured one of them did not have friendly relations with the US." That is, they chose a country that would not share intelligence with the United States that might have revealed their operations to US authorities.

The loss of information was enormous and wide-ranging. A May 3, 2012, email contained an attachment that appeared to lay out the entire flight test plan for the F-35. The 630,000 stolen files related to the C-17 covered virtually every aspect of the plane's design.

"The drawings include the aircraft front, middle and back; wings; horizontal stabilizer; rudder; and engine pylon. The contents include assembly drawings, parts and spare parts. Some of the drawings contain measurement and allowance, as well as details of different pipelines, electric cable wiring, and equipment installation," wrote UC1. "Additionally, there were flight test documents."

In other words, they had stolen much of what one needs to build and fly a plane. They concluded, again with a flourish, "This reconnaissance job, because of the sufficient preparations, meticulous planning, has accrued rich experience for our work in future. We are confident and able . . . to complete new mission."[14]

Given their tendency to play up their successes, the FBI granted in the criminal complaint, "The success and scope of the operation could have been exaggerated."

However, many of the files they claim to have stolen exactly match the genuine files supplied to the FBI by the target companies as part of the bureau's investigation. And as US intelligence officials often note with bitter sarcasm, China's Y-20 transport aircraft looks awfully similar to Boeing's C-17—just as China's J-31 fighter looks almost indistinguishable from the F-35.

Su's team achieved remarkable success over an extended period of time, through a combination of human collection and cyber intrusion. And they were driven by a combination of motives as well: money and country. Bob Anderson says those goals are not mutually exclusive. In fact, they work just fine as motivators in tandem.

"The thing with the Chinese, it goes both ways all the time," Anderson told me. "People want to make money. I take care of you, you take care of me. But at the same time, they're very cognizant of what their nation is asking them to acquire.

"The subtle difference is they all also want to be compensated," Anderson continued. "Not necessarily from the government, but any kind of backdoor deal that they can make on the side that gets the money."

That is not a combination US law enforcement sees with America's other biggest cyber adversary, Russia.

"You'll never see them sell that back to Russia and then try to sell that someplace else. Never happen. It'll never happen," said Anderson. "You'll never see that with the SVR, FSB, GRU [Russia's foreign, domestic, and military intelligence agencies], whoever from Russia.

"[For] China, it's almost a way of doing business," Anderson concluded.

Their email correspondence demonstrates that money was never far from the minds of Su Bin and his partners. Many of the exchanges were plaintive, even petty. In a 2010 email, the team notes that they had far outspent the RMB 2.2 million in funding they had received to that point to "build up its team and infrastructure." They claimed their actual expenses were more than triple the figure: RMB 6.8 million, or nearly $1 million in US dollars. They had to cover the difference with a bank loan. This budget shortfall, they said, forced them to "miss the best opportunity" for information on the C-17.

All three team members were frequently asking for updates on their compensation. On March 30, 2010, one of the China-based hackers emailed Su to ask if he "had any good news," presumably regarding payment. Six days later, on April 5, he emailed Su again with the subject line, simply, ". . . ."

To jump-start the payment process, he offered to share a sample of the stolen C-17 material with their buyers to pique their interest. The same day, Su emailed back to reject the idea, advising his partner, "If money is collected for the sample of 17, it won't be easy to collect your big money that would follow."

Commiserating, he added a swipe at Chinese government bureaucracy: "Also, it's a long process to apply for the expenses."[15]

Su, however, did his own grumbling. In March 2010, he wrote his partners to complain about the state-owned aircraft manufacturer that was buying the C-17 data: "This information is what the [Chinese aircraft corporation] needs. They are too stingy!"

In Anderson's view, the desire for financial reward does not weaken the commitment of China's information thieves. Nationalism remains a powerful motivator—one that inspires many Chinese citizens to do their part for their country, whether they are formal government operatives or part-time volunteers.

"The MSS [Ministry of State Security, China's foreign and domestic intelligence service] [and] the [Chinese Communist] Party in China look at every individual in the mainland as a collector.

And their family, extended family as a collector," said Anderson. "Their strength is their people."

This is an advantage, Anderson says, that is unique to China.

"That's even different than the SVR [Russia's foreign intelligence service], FSB [Russia's domestic intelligence service] from Russia. That's different than the NIS [South Korea's intelligence service]. That's different from the SIS [the United Kingdom's foreign intelligence service, known as MI6]," he continued. "That's the strength, I think, that China has when it comes to this."

———————

Su Bin's run as one of the most damaging Chinese spies of his generation would end in the summer of 2014, when he was arrested in Canada on a US-issued warrant five years after he sent his first instructions to his co-conspirators in China detailing targets for hacking inside the United States. A Justice Department statement announcing his indictment said he had "worked with two unindicted co-conspirators based in China to infiltrate computer systems and obtain confidential information about military programs, including the C-17 transport aircraft, the F-22 fighter jet, and the F-35 fighter jet."[16]

Two years later, in February 2016, Su consented to be sent to the United States, where he pleaded guilty before US district court judge Christina Snyder in California. In his plea agreement, the Justice Department announcement said, "Su admitted to conspiring with two persons in China from October 2008 to March 2014 to gain unauthorized access to protected computer networks in the United States, including computers belonging to the Boeing Company in Orange County, California, to obtain sensitive military information and to export that information illegally from the United States to China."[17]

Su also admitted to seeking financial profit from the theft. In July 2016, he was sentenced to forty-six months in prison and ordered to pay a ten-thousand-dollar fine.

In announcing his plea, then–assistant attorney general John P. Carlin said, "This plea sends a strong message that stealing from the United States and our companies has a significant cost; we can and will find these criminals and bring them to justice. The National Security Division remains sharply focused on disrupting cyber threats to the national security, and we will continue to be relentless in our pursuit of those who seek to undermine our security."

"Cyber security is a top priority not only for the FBI but the entire U.S. government," then–assistant director of the FBI's Cyber Division James Trainor added. "Our greatest strength is when we harness our capabilities to work together, and today's guilty plea demonstrates this. Our adversaries' capabilities are constantly evolving, and we will remain vigilant in combating the cyber threat."[18]

They were bold statements, fitting with the seriousness with which the United States takes such cyber theft and which it hopes to make clear to its adversaries. And issuing charges against foreign hackers, even when the United States has no ability to arrest them, is one part of the US response to cyberattacks: that is, naming and shaming the perpetrators. "It sends a message," US law enforcement officials often say. But beyond the embarrassment of being caught and the damage to future intelligence collection from what the United States learns about Chinese cyber tradecraft, such charges have not changed Beijing's behavior.

"We're never gonna see those guys," Anderson noted. "And the Chinese would come right back and tell us, 'One, we deny everything, and two, we don't care. We don't even honor what an economic espionage statute is. We don't even agree that it exists.'"

Su Bin's indictment in 2014 was a victory for the FBI and an example of good cyber police work. Despite Su and his team's prodigious efforts to cover their electronic tracks, FBI analysts successfully followed their electronic trail across the globe, through multiple countries and multiple "hop points," and traced it all back to one friendly and gregarious Chinese businessman.

As with many espionage cases, there were hints and clues of operations targeting Boeing and other US defense contractors before Su Bin and his team were identified and caught. Bob Anderson remembers seeing intelligence in the months prior to his arrest indicating a possible breach.

"In a lot of [cyber] cases, no different from espionage cases, you'll get a whisper or a bit of intelligence that says there's a guy," said Anderson.

Identifying that "guy," however, took years in this case, allowing China to advance its military technology by years or more in the process. And today China is flying two jets that at least look almost exactly like the F-35 and C-17.

Anderson and others warn that Su's team was just one node in a massive global web of Chinese cyber spies.

"There's a lot of him out there. I think what people need to understand, this isn't a one-off, and it's not gonna stop just because you indict one guy, or you indict hundreds," Anderson warned. "They don't slow down one iota.

"There's hundreds if not thousands of these people in my opinion that are here or in our friendly partner countries because now it's a global business environment," he added.

More alarmingly, Anderson warns that the FBI's Cyber Division is aware of, perhaps, 10 percent or less of all cyber intrusions like the one carried out by Su Bin and his partners. They are simply overwhelmed and often overmatched.

This battlefield of the Shadow War is constantly changing. Just as the United States catches one or several spies, new ones enter the fray with new weapons. The technology is constantly advancing and therefore so are Chinese hacking tactics. Already the modus operandi Su Bin and his team employed so successfully—that of having an operator on the ground in the States identify targets for operators based in China to hack—is a thing of the past. China's cyber capabilities have advanced to the point that Beijing no longer needs an operator on the ground in the United States. More and more, both the targeting and the hacking can be done from afar, within the safe confines of China.

"I don't think they need it anywhere near what they needed five years ago because of cyber, because of different aspects to get into very highly restricted areas of companies, because of their ability to hack and their ability to get quieter than they used to be," Anderson told me.

"The one thing China is very good at is continually modifying their tradecraft," he added.

That means that what the FBI learned from the Su Bin case may not help them catch the next Su Bin. They are looking for a different kind of thief with a different MO.

"If you're looking for Su Bin now you're gonna miss him. They're already out here. That's a key point: if you're looking for him, you've lost it."

For the US military, the extent of the damage from the hacking by Su Bin and his partners is not entirely clear. China has since deployed similar aircraft with similar capabilities. However, US military officials have told me, often with some derision, that China's J-31 fighter and Y-20 transport are, at best, cheap facsimiles.

Bob Anderson is less sanguine. He is not a military commander. He has spent his entire professional life in law enforcement. However, he has seen the intelligence. And when I asked him how much sensitive data Su Bin and his team had stolen relating to some of America's most advanced military aircraft, his answers were disturbing.

For the C-17 he said, simply, "A lot, a lot." For the F-35, he went a little further: "A lot, enough to where I think it's a huge problem."

China, in the span of five years and with just three operatives, had at least narrowed the gap with the United States on three of its most advanced military aircraft—aircraft that had taken more than a decade for the United States to develop and tens of billions of dollars to design and manufacture. And China had done so not just with the intention of catching up to the United States technologically but also to level the playing field with the Americans in the event of war—a prospect some military commanders on both sides see as inevitable.

Anderson and other intelligence and law enforcement officials I've interviewed speak of China with a spy's grudging respect for their adversary.

"To know them is to respect them, if you're ever gonna understand what they're doing," said Anderson. "You might not like them. You might not agree with what they do. But you better respect how they do it or they're gonna get over on you every single time, because they're very good at what they do.

"We are looked at as the most significant adversary they've got and they're gonna lie, cheat, and steal . . . to figure out how they're gonna get ahead of us," Anderson told me. "I don't think people look at it that way."

LESSONS

Su Bin's enormously successful conspiracy to steal secrets detailing some of the US military's most advanced aircraft contains two sobering lessons for the United States. First, China has been aggressively stealing US government and private sector secrets and intellectual property for decades. For China, this state-sponsored theft is not a crime, but a policy. Its cost to the United States is estimated in the tens of billions of dollars per year. By that measure, it is arguably the most expansive theft in modern history—and it is ongoing. For China, the goal of this theft is not just to catch up to the United States, but to surpass it. And that intention is not a secret, kept under wraps in the deep recesses of Beijing's Zhongnanhai government compound, but one that Chinese officials discuss openly in speeches and official publications.

Second, repeated US efforts by multiple administrations of both parties to defend against, and warn Beijing away from, such attacks have failed to change Chinese behavior. This includes everything from President Obama's personal warning to Chinese president Xi Jinping, to the US Department of Justice indicting members of the Chinese military, to President Trump imposing hundreds of billions of dollars in tariffs on Chinese imports. Moreover, similar to the West's approach to Russia, US leaders and policy makers mistakenly persisted in the view that China wants what the West wants, that is, accession to a rules-based international order. To compound the error, US government and business leaders boldly proclaimed that China's entry into international treaties and associations, such as the World Trade Organization, would change its behavior over time. In fact, its entry likely facilitated rather than curtailed China's theft of America's greatest asset: its intellectual property. It's a lesson the West is still learning today.

CHAPTER 4

Little Green Men

(RUSSIA)

In the early afternoon of July 17, 2014, Alexander Hug was in his office in Kiev, Ukraine, when he received the first reports of a downed aircraft. As a leader of the Special Monitoring Mission to Ukraine for the Organization for Security and Co-operation in Europe (OSCE), he led the dangerous and often frustrating task of documenting the bloody shooting war between the Ukrainian military and pro-Russian separatists armed, funded, and directed by Russia. He and his team monitored the ever-changing front lines, attempted to facilitate dialogue, and—most troublingly—kept track of the growing toll of civilian casualties. Activity in the air only drew their attention when a jet went down.

In the days and weeks prior to that hot July afternoon, Hug's team had been seeing more frequent media reports of downed military aircraft. Ukraine maintained a modern and sizable air force that, for a time, commanded the airspace over the battlefield.

However, as losses among Russian-backed fighters mounted, Moscow had begun supplying them with shoulder-fired antiaircraft missiles, or MANPADs (for "man portable air defense system"). The decision would have an immediate and devastating impact.

Just over a month earlier, on June 14, Russian-backed forces had brought down a Ukrainian IL-76 military transport jet on approach to the eastern city of Luhansk, killing all forty-nine people on board. The next day, as wreckage and bodies littered the farm fields outside the town of Novohannivka, a commander of separatist forces gleefully claimed credit for the jet's loss. In a video posted on YouTube, Valery Bolotov said, "I can't tell you anything more detailed on the IL-76 but I will repeat [it] was hit by our militia, the air defense forces of the Luhansk People's Republic." His forces had used a shoulder-fired rocket, manufactured and supplied by Russia. Since the jet was on approach, it was at an altitude easily within range of a shoulder-fired weapon.

More recently, however, Hug noticed news accounts of aircraft shot down at much higher altitudes, far beyond the range of MANPADs. On July 14, just three days earlier, another Ukrainian military transport, an Antonov-26, was shot down near Luhansk at an altitude of 6,200–6,500 meters, nearly 20,000 feet.[1] Two days later, on July 16, a Ukrainian Sukhoi Su-25 fighter jet was shot down at 6,250 meters.[2] Striking aircraft at that altitude required more powerful surface-to-air missile systems with much greater range than shoulder-fired weapons—range that extended up into the cruising altitude of civilian aircraft over Europe.

"At the time, there were several reports of aircraft being shot down, including the day before, so we of course took note," Hug remembered. "But shortly after, the first messages came in that this one may have been a civilian aircraft."

To this day, Hug still keeps the first notes he jotted down that afternoon on an easel-sized flip chart in his office in Kiev. He read them back to me four years later: "Boeing 777–200," "Amsterdam to Kuala Lumpur," "last contact at 1630," "approximately 300 passengers."

"These were my first scribbles," Hug told me, his voice growing muted as he remembered. They were also the first signs that he and his team were facing something they hadn't prepared for.

Four years later, the memories are clearly still raw for Hug. Soft-spoken despite his height, Hug has a calm demeanor that seems to be in constant battle with all he has witnessed in his role documenting a bloody war in modern Europe. As a father of three, Hug says that his children are never far from his mind as he chronicles the human cost of the fighting. The victims of this crash would do the same for him and in even starker, more painful terms.

With its origins in the Cold War, the OSCE had seemed outdated in post–Cold War Europe. Founded in 1973 by US president Richard Nixon and Soviet Communist Party general secretary Leonid Brezhnev, the OSCE had at one time monitored the signature nuclear arms agreements of the 1970s and '80s.[3] After the collapse of the Soviet Union in 1991, however, the OSCE's importance and profile faded. It took on new, less ambitious missions, such as monitoring postwar Kosovo and Bosnia in the mid-1990s. Its once lofty role as peacekeeper between the superpowers seemed to have passed. In 2014, however, Russia's invasions of Crimea and Eastern Ukraine were thrusting the group back into the international spotlight.

On July 17, Malaysia Airlines flight MH17 had left Amsterdam's Schiphol Airport bound for Kuala Lumpur, Malaysia, at 12:31 p.m. Dutch time (10:31 a.m. GMT) with 283 passengers and fifteen crew on board.[4] The Boeing 777 reached its cruising altitude of 33,000 feet heading southeast over Germany, Poland, and then Ukraine.

The weather was clear, the flight was smooth, and the broad plains of Eastern Europe were unfolding outside the passengers' windows.

Near the time of MH17's departure from Amsterdam, a resident in Makeevka in Eastern Ukraine snapped a photo of a truck carrying a Russian-made BUK-TELAR surface-to-air missile through the town. Makeevka lay right in MH17's flight path, some three hours ahead.

In the summer of 2014, Malaysia's national airline was flying under the cloud of the unexplained loss of another Boeing 777—the notorious flight MH370, which had disappeared four months earlier en route from Kuala Lumpur to Beijing. A massive, international search effort had still not located any sign of the plane—and the cause of its disappearance remained a mystery. As he waited to board MH17 in Amsterdam, Cor Pan, a Dutch citizen like the majority of his fellow passengers, posted a picture of the jet on his Facebook page, joking: "If the plane disappears, this is what it looks like."

Two hours and forty-eight minutes into the flight, at 3:19:56 p.m. Dutch time (13:19:56 GMT), MH17 was cruising over Eastern Ukraine, near the Russian border, when regional air traffic control communicated permission to the flight crew to continue their journey into Russian airspace. MH17's crew acknowledged the transmission.[5]

According to the investigative report by the Dutch Safety Board, four seconds later, at 3:20:00 p.m. Dutch time (13:20:00 GMT), air traffic control radioed the cockpit again with further instructions. It received no response. The control tower made four further attempts to reach the crew, asking each time, "Malaysian one-seven, how do you read me?" Again, it received no answer.

Growing concerned, the controller in Europe radioed his counterpart across the border in Russia, asking, "Rostov, do you observe the Malaysian seventeen [MH17] by the transponder?"

Without communications with the cockpit, air traffic control wanted to know if the signal from the jet's transponder was still being detected. Rostov responded, "No, it seems that its target started falling apart."[6]

The cockpit voice recorder, later recovered at the scene, would show that the recording stopped abruptly at 13:20:03 GMT, seven seconds after its final transmission. Again, according to the Dutch investigative report, a high-energy sound wave lasting 2.3 milliseconds was detected just before the end of transmission. Acoustic analysis would find that the sound originated from outside the plane, above the left-hand side of the cockpit.

Within minutes of its final transmission, residents in and around the Eastern Ukrainian town of Grabovo, thirty miles northeast of Makeevka, began sharing videos and pictures on social media of plane wreckage. Eyewitnesses described an explosion and giant fireball in the sky, followed by pieces of burning debris crashing down to earth.

As those social media reports made their way back to Kiev, Hug noted that their location was in the epicenter of the area of armed conflict at the time. The site was a long way from the capital, nearly five hundred miles by car. Hug instructed his team to pack and be ready to go in the morning.

On Ukraine's uneven and aging roads, the trip would have taken a full day by car. By that time, more than twenty-four hours after the crash, Hug worried that bodies and key evidence might already be removed by residents or, more worryingly, removed or tampered with by pro-Russian forces on the ground. His most immediate concern, however, was his team's safety. They would be attempting to establish the facts of a deadly crash in the middle of an active war zone.

Hug found one local helicopter charter company willing to

fly them in, free of charge. By now word of the crash had spread widely across the country. Bloodshed had become all too familiar in Ukraine in recent months—and average citizens were offering to help.

The OSCE monitoring mission was not a peacekeeping force. None of its members were armed. They were well-intentioned bureaucrats attempting to keep honest eyes on a conflict mired in lies and deception. Hug felt responsible for his whole team's safety. And if a passenger jet streaming through the sky at six hundred miles per hour six miles above the earth had been at risk, a slow-moving helicopter flying a few hundred feet off the ground was an easy target. It would be a nervous trip.

"I trusted these pilots that they knew what they were doing, but the conflict was very fluid at the time, so there probably was some risk," he said, with dry understatement.

Early in the morning of July 18, they landed safely in an open field south of a town called Izium. From there they navigated south toward the crash site in armored vehicles arranged by the OSCE headquarters in Kiev. As they drove the remaining miles, they could see a plume of smoke rising in the distance. They were close, within sight of wreckage that was still burning, when they were met by an unexpected obstacle.

The soldiers they encountered wore no insignia on their camouflage uniforms. Hug immediately recalled the "little green men" who had showed up on the streets of Crimea four months earlier, speaking Russian and carrying Russian weapons but identified by the Russian government only as concerned citizens "volunteering" to protect Crimea's ethnic Russian population. It wasn't so much a disguise as a cynical prank. The accents and weapons of the soldiers confronting them gave them away as foreigners. And Hug himself would later interview captured separatist fighters who admitted they belonged to regular Russian military units deployed to Ukraine by Moscow. But at the time, without flags and name tags, who could prove where these soldiers were from?

Now, in Eastern Ukraine, another contingent of "little green men" had appeared. At the crash site, the leader of the unit was a bulky Russian donning camoflauge fatigues and hat, who struck a dramatic pose in front of the OSCE convoy.

"He stood in the middle of the road with a machine gun—not just an AK-47, it was a machine gun—and he wouldn't let us further than two or three hundred meters down the road that ran parallel to the crash site," recalled Hug.

"He was obviously intoxicated. I could smell his breath," Hug added. "He's drunk, I thought."

Hug and his team were trapped in a bizarre standoff, within yards of the deadly crash they had been sent to document but with no freedom to do their work. As the OSCE monitors were held at bay, local emergency workers, Russian soldiers, and even a handful of journalists were moving freely around a crash site strewn with bodies, body parts, and still-smoldering pieces of wreckage. This was very possibly an international crime scene, but it was not being treated as one.

"It was still quite a wild mess in that area. There was no order," said Hug. "Armed men, journalists, and other civilians were still roaming around in the fields of debris."

This "wild mess" was right on the contact line between Russian and Ukrainian forces. Casualties had been rising on both sides in the previous weeks. Neither side was willing to let down its guard, despite the shattered aircraft now littering the battlefield. The scene was tense and dangerous. Hug could hear heavy artillery in the distance. The armed man in charge fired his rifle in the air when a member of Hug's team tried to walk closer to the wreckage.

"We tried to move on but there were shots fired over our heads by the armed men there to push us back, and to make clear, they would not let us go any further," he said.

Residents in the area were finding bodies in the streets, in their gardens, and on their rooftops, and taking pictures to show to Hug and his team. That first day, Hug observed at least twenty-one bodies, some of which were already showing early signs of decomposition in the oppressive heat. This was the height of summer. Temperatures were soaring above one hundred degrees Fahrenheit.

"The bodies were marked but left exposed to the elements," Hug recalled. "A uniformed rescue team present at the scene informed me and my colleagues that their task was to mark the bodies but not to remove them."

Pressed repeatedly by Hug, no one could say whose responsibility it was to remove the bodies. Local residents were overcome.

"They were all emotionally overloaded. Some of them were crying, some of them trying to help as well," said Hug. "The civilian population had this double trouble of coping with this incident while being shelled and fired upon."

Later in the day, the Russian commander allowed Hug's team to venture two hundred yards into the crash scene but no farther. The immense toll was unescapable. They saw a part of the plane's tail section and other debris, including passenger seats and baggage containers. Residents had begun to make piles of luggage, out of respect for the victims. They were forced to leave—at gunpoint—after just seventy-five minutes.

Hug was not instructed to determine the cause of the crash. However, from the beginning, he spotted details that were potentially revealing. The outer skin of the plane, especially around the cockpit, was pockmarked with jagged holes. Hug noticed the edges of the holes were turned inward, indicating projectiles had slammed into the plane from the exterior at high speed. He had served in the military and recognized the blast pattern from explosive ordnance.

"I saw different parts that had been punctured, and from my military background, I could tell this was a shrapnel-type of impact," Hug recalled. "I could also tell it must have been at high velocity."

––––––––––

Hug was not alone in his suspicions. The day of the crash, more than five thousand miles away, a different team of investigators was taking its first look at the disappearance of MH17. At the very moment MH17 had lost contact with air traffic control, US surveillance satellites had captured a flash in the sky over Eastern Europe. And at the Missile and Space Intelligence Center (MSIC) of the Defense Intelligence Agency (DIA) in Huntsville, Alabama, US intelligence analysts were forming a team to examine the satellite data to determine what had caused that flash. I was the first reporter to be allowed inside the technical analysis room of the MSIC, where I met members of the investigative team who had been on duty that July morning. Here the technicians say they perform "CSI" for war zones.

"It is the CSI forensic sort of capability, similar to crime scene investigation sort of thing, a little bit of DNA here and a fingerprint there begins to piece together a pretty compelling story," said Randy Jones, the chief scientist at MSIC. "All those things provide little pieces of the puzzle to fit together to give you the picture of what actually happened. So, this room is where the puzzle gets put together."

For the analysts at MSIC, the pieces of this puzzle were drawn from a massive amount of satellite and radar data. Huntsville, Alabama, has a long and storied history in missile technology. The city is decorated with examples of the most powerful rockets ever made, towering monuments to America's nuclear and space programs. The home of the US missile program since the 1940s, it still boasts a number of German restaurants, a legacy of the Nazi Germany exiles, led by Wernher von Braun, who helped jump-start the US rocket and missile program following World War II.

As the nuclear age descended on the United States and the world, the DIA's missile intelligence became focused on identifying and tracking incoming missiles as well as launching them into the sky. Today at the DIA's disposal is an entire constellation of satellites orbiting the earth at some 22,000 miles up. Together with vast arrays of land-based radar systems, they are the nation's early-warning system for nuclear attack—designed to detect any explosion across the globe that could indicate the launch of a missile toward the US homeland.

To help distinguish between hostile and friendly launches, DIA analysts maintain a deep knowledge of every foreign missile system. In their backyard in Huntsville, they have parked a collection of some of the most notorious foreign missiles, bought on the international weapons market or "acquired" from adversaries by other means. They will not specify exactly how. I was able to climb up onto the mobile launcher of a SCUD like the ones Saddam Hussein fired at Israel and Kuwait during the Persian Gulf War, press a button, and raise the nearly forty-foot-long behemoth. The SCUD, however, is decades-old technology.

Today, as Russia, China, North Korea, and Iran constantly upgrade their missile programs and expertise, the DIA is constantly upgrading their ability to track them. Analysts assured me they have studied and made preparations for every conceivable missile threat. Among them is Russia's SA-11, or BUK-TELAR, surface-to-air missile system.

"Not just can I tell what weapon system it is, but I can tell what it's done, what it's doing, what it's about to do, from the patterns that they see in that data," Jones said with obvious pride.

On that July day, they took on a new and unfamiliar mission: attempting to identify the cause and the culprit behind the loss of a passenger jet over Europe. The questions for the intelligence analysts were straightforward: Was there an explosion at the time contact was lost? Did it come from inside or outside the plane? Inside could indicate a terrorist attack. Outside could mean a missile. And, if the explosion did come from outside, was there satellite evidence of a missile launch just before it?

As luck would have it, they had visitors that day: a group of representatives from across the intelligence community whose expertise was in just this kind of analysis.

"We had them here in the building, reviewing activities from a year's worth of adversary threats," said Jones. "It just happened timing-wise to work out that way."

The DIA team went to work immediately after hearing the first reports of the crash. Within an hour they had assembled a team of intelligence analysts assigned to MH17. Within an hour and a half, the team had pulled together all the relevant satellite and radar data. Using the satellite and radar data and precise timing, they were able to establish a trajectory consistent with a surface-to-air missile.

"Within an hour and a half, we were confident that it was a missile that shot it down," Jones told me later. "A surface-to-air missile that shot it down. We had a fair idea which one, although we still had some homework to do."

Crucially, their analysis identified the missile launch site: a farm field near the town of Pervomaiskyi—inside territory then controlled by pro-Russian forces.

To eliminate any doubt, pro-Russian separatists once again bragged about their kill. Just thirty minutes after the crash, a separatist commander posted a video on social media, which Secretary of State John Kerry cited in a July 20 interview with Fox News.[7]

"We know that the so-called defense minister of the People's Republic of Donetsk, Mr. Igor Strelkov, actually posted a bragging social media posting of having shot down a military transport," Kerry said. "And then when it became apparent it was civilian, they pulled it down from social media."

Strelkov was a well-known figure among the separatists and, like many of them, a veteran of the Russian military. The post went up on Strelkov's page on a Russian version of Facebook at 5:50 p.m. local time, about a half hour after MH17 disappeared. Attached were videos similar to eyewitness videos of the MH17 crash. Strelkov's post read: "In the region of Torez AN-26 [an Antonov 26 military transport] plane has been shot, it is somewhere near the 'Progress' mine. We have warned them—not to fly 'in our sky.' Here is video-proof of yet another 'bird fall.' The bird has fallen behind the [waste heap], it missed the residential quarters.[8]

"Peaceful citizens were not hurt," he claimed.

According to US intelligence officials, Mr. Strelkov's social media post was corroborated by other communications among separatists intercepted by US and Western intelligence agencies. The evidence indicated that the pro-Russian separatists had committed a horrifying error, mistaking a passenger jet over Europe for a Ukrainian military aircraft, specifically an Antonov-26, or AN-26, like the one they had shot down three days earlier. Back at the DIA, the evidence was seen as conclusive.

Images of the wreckage taken at the scene would provide more evidence to confirm their assessment. What the analysts saw in those photos was consistent with the explosive pattern of an SA-11.

"So, we looked at the fragment patterns here, the density of the fragment, fragment holes, and where that happened on the aircraft," Jones explained. "And then our modeling of how the SA-11 flies, we assessed that the warhead went off about twenty feet away from the aircraft, to the top left of the cockpit."

SA-11s are designed to explode just in front of and above their target, to maximize damage from the burst of shrapnel. The missile had streamed up from the ground, outside the pilots' field of vision and at a velocity too high to register with the naked eye. The jet was torn apart without warning.

"By that afternoon, we published a report saying that we assessed that MH17 was shot down by an SA-11, fired from separatist cell territory within Eastern Ukraine," Jones told me.

That evening, US intelligence agencies delivered their assessment to the White House, informing President Obama that Russian-backed separatists had destroyed a commercial aircraft over Europe with a powerful missile supplied by Russia. All 298 passengers and crew were dead.

"So that same day, less than twelve hours, we had a high-confidence assessment," Jones told me. "In the intelligence community, high confidence carries some weight with it that you have pretty convincing evidence.

"It was a mixed-emotions day. It was a day when your job takes a little new significance and there's a certain urgency to finding out what happened. But then the actual casualties and all bring a sobering side to it," Jones told me.

Within twenty-four hours, US intelligence agencies were confident in both the cause and the culprit of MH17's loss: a Russian-made missile fired by pro-Russian separatists. US officials and policy experts inside the government were privy to that assessment. However, some were still expressing caution and doubts.

On the day MH17 went down, Ambassador Geoffrey Pyatt was nearly a year into his term as the top US diplomat in Ukraine. Pyatt was a foreign service officer of twenty-five years and veteran of postings in Honduras, India, and Vienna with international organizations including the International Atomic Energy Agency (IAEA). None of his postings were as difficult and sometimes dangerous as ambassador to a country at war with Russia.

The day after the MH17 crash, Pyatt recalls a contentious videoconference with Obama administration officials in Washington.

"That was one of the darkest days of my time in Ukraine," said Ambassador Pyatt. "I remember one of my Washington colleagues saying something along the lines of 'We have to be very careful not to jump to conclusions.'"

That answer was too much for Ambassador Pyatt to bear.

"It was one of my more unguarded moments because I remember saying very clearly, 'You say we don't know what happened, but we do know. We do know that Russia is responsible, that there were no Ukrainian missiles of this class in the region, and one way or another, the Kremlin is responsible for the deaths of three hundred people.'"

The loss of MH17 would finally spark a sea change in the way the United States and Europe viewed Russia.

"It catalyzed the Europeans to come together on the kind of tough sanctions response that we had been pushing for weeks and weeks," said Pyatt.

However, the fact that it took mass murder in the skies over Europe to generate such a response demonstrated once again how much the United States and the West underestimated Russia and persisted in assigning false limits to Russian ambitions and aggression. Russia was deep into the Shadow War, long before US and European leaders would recognize it. And it would take the deaths of the 298 passengers and crew of MH17 to dismantle the West's misconceptions and spark more decisive action.

R ussia had, in fact, made its plans for Ukraine clear nearly ten months earlier. In September 2013, a select group of the world's power brokers gathered for a conference at the elegant and imposing Livadia Palace in Yalta, Crimea. The palace drips with history beyond its neoclassical and Moorish architecture. Nearly seventy years earlier, Franklin Roosevelt, Joseph Stalin, and Winston Churchill met on the same grounds to decide the fate of post–World War II Europe and delineate the boundaries and balance of power between the West and the Soviet Union for decades to come.[9]

The 2013 conference, which was organized and sponsored by the Ukrainian billionaire Victor Pinchuk, included its own impressive list of world leaders and corporate power brokers: Bill and Hillary Clinton, Tony Blair, David Petraeus, Bill Richardson, then–president of Ukraine Viktor Yanukovych and his soon-to-be successor, Petro Poroshenko, Gerhard Schröder, Dominique Strauss-Kahn, as well as a close Putin advisor, Sergei Glazyev, who would deliver the conference's most memorable speech.[10]

The dominant topic of conversation, while not quite deciding the fate of the Western world, carried its own weight in Europe. Ukraine was then negotiating a free trade and political association agreement with the European Union, an arrangement vehemently opposed by its neighbor Russia. Still, the mood was largely positive. Few expected Russia to stand in the way. A more integrated Ukraine, they thought, was in both Ukraine's and Russia's economic interests.

Crucially, US diplomats were careful not to venture into talk of Ukraine's potential admission into NATO, a far more direct challenge to Russia's national security interests. In fact, US diplomats were quietly urging their Ukrainian counterparts not to overreach by venturing into territory that would inevitably be seen as provocative by the Kremlin.

Glazyev's message, however, was more expansive. He made clear that any Ukrainian partnership with Europe would be a mistake and dismissed any promised or perceived benefits of a Ukrainian partnership with Europe as "mythology."

"Who will pay for Ukraine's default, which will become inevitable? Would Europe take responsibility for that?" he asked.

He went further, issuing a threat to Ukrainian leaders, which elicited jeers from the crowd.

"We don't want to use any kind of blackmail," he warned. "But legally, signing this agreement about association with the EU, the Ukrainian government violates the treaty on strategic partnership and friendship with Russia."[11]

Ambassador Geoffrey Pyatt had just begun his term as the US envoy to Ukraine.

"It was extremely confrontational. He was speaking to the Ukrainians and he basically said if you persist in this task or this course of association with the European Union, if you pursue a free trade agreement with the European Union, if you pursue a closer relationship with the Europen Union, we will make it extremely painful for you and very bad things are going to happen to Ukraine and to the people of Ukraine," said Pyatt.

At the time, Pyatt recalls himself and many others in attendance dismissing the Russian threats. The momentum for Ukraine's greater integration into Europe was strong, both in European capitals and among Ukrainians themselves.

"I don't think anybody took those warnings seriously enough," says Pyatt. "I think it reflected a strategic error of judgment that we and the Europeans made."

Not everyone in attendance was so sanguine. Pyatt recalls a quiet meeting with one European diplomat who seemed to grasp the degree of Russian anger at Ukraine's drift westward. The diplomat was, interestingly enough, the EU official whose job it was to grow the union's membership, the so-called Commissioner for Enlargement, Stefan Fule.

"I sat down with Fule over a table like this. It was just a small informal bilateral. It was the first time I ever met him, and he came on very, very strongly and said basically, 'Where the hell are the Americans? Don't you realize there is a great struggle that's going on right now to define the future of the European periphery? We need an engaged America.'"

Failing to hear the brash and confrontational message behind Glazyev's speech at Yalta would prove to be another in a series of missed signals and warning signs from the Kremlin regarding Ukraine. Over the next several months, US and European diplomats and policy makers would persist in mirror-imaging their Russian counterparts, while Putin and his lieutenants were playing by very different rules.

Russia had sent warning signs even years before Yalta in 2013. Many Russia experts now point to the speech by Russian president Vladimir Putin six years earlier, in February 2007, at the Munich Security Conference. There Putin shocked attendees including US defense secretary Robert Gates and soon-to-be presidential candidate Senator John McCain with a searing diatribe against US foreign policy.

"One state and, of course, first and foremost the United States, has overstepped its national borders in every way," he said.[12]

The speech heralded the end of the Bush administration's efforts to cultivate a warmer, more cooperative US relationship with Russia, as Putin took aim at US military action abroad, including declaring the Iraq War "illegal."

"This is very dangerous. Nobody feels secure anymore because nobody can hide behind international law," he said. "We're witnessing the untrammeled use of the military in international affairs. Why is it necessary to bomb and to shoot at every opportunity?"[13]

Beyond condemning US military action abroad, he questioned the usefulness of international arms treaties that had laid the groundwork for peace between the superpowers for decades, including the 1988 US-Soviet agreement to ban intermediate-range ballistic missiles. His anger was sparked by a recent US decision to deploy an antimissile defense in Europe to protect against the growing Iran nuclear and missiles threat.

"I don't want to suspect anyone of aggressiveness," Putin charged. "But if the antimissile defense is not targeted at us, then our new missiles will not be directed at you."

Today, Putin's 2007 speech reads like an action plan for a new Russian foreign policy that would redefine Russia's relations with the West in the decade to come. And in fact, Russia fired an early warning shot just two months after Munich, launching the devastating cyberattacks on Estonia beginning in April 2007. However, successive US presidents and secretaries of state would continue to attempt to improve—to "reset"—Washington's relations with Moscow, to little effect. Two years later, the Obama administration, led by Secretary of State Hillary Clinton, unveiled the famous "reset button" alongside her Russian counterpart Sergei Lavrov in Geneva. As a candidate and as president, Donald Trump would speak repeatedly of building a "good relationship" with Russia, including holding a summit with President Putin in Helsinki in July

2018, even as Russia amplified its attacks. Two thousand thirteen into 2014 would prove to be a particularly pivotal period and a turning point for the worse.

R ussia would begin to turn its threats into more violent action in Ukraine in late 2013, demonstrating the West's consistent misreading of Moscow and Moscow's often paranoid misreading of the West.

Two months after the Yalta conference, in November 2013, the Ukrainian government was preparing to sign the European Union–Ukraine Association Agreement. Most Ukrainians welcomed the prospect of a closer relationship with Europe. Moscow did not. With the Kremlin exerting pressure on Ukrainian president Viktor Yanukovych, a close ally of Putin, signs began to emerge that the agreement would not be signed. On November 21, by decree, the Cabinet of Ministers suspended the preparations. Ukrainian citizens were outraged. By that evening, the first protesters began to appear on Kiev's Independence Square, better known as the Maidan, which means simply "square" in Ukrainian.

That first night, the crowds were modest, but by November 24 they had grown enormously. The first major pro-European demonstration of what would come to be known as "the Maidan" after the name of the square where they met drew some fifty to a hundred thousand people. Their demands were ambitious: no longer just signing the Association Agreement, but now the resignation of Ukraine's pro-Russian government and dissolution of parliament. Four days later, at the EU summit in Lithuania, Ukraine's government formally rejected the Cooperation Agreement. The protesters—outraged further—made clear they were in the Maidan to stay, erecting tents on the square. By November 30, a core group of two hundred to a thousand protesters was maintaining a round-the-clock presence.[14]

This was a turning point for the demonstrators and for the Ukrainian government. According to a report later conducted by Ukraine's Prosecutor General's Office, the country's top national security and interior ministry officials decided that night to disperse the protesters by force.[15]

"A smart strategy for Yanukovych would have been to just let this thing burn itself out," said Ambassador Pyatt. "Because it was going to get colder and colder and colder. And for a while, it looked like that was going to be the strategy."

Early in the morning of November 30, according to a later report by Human Rights Watch (HRW), "Riot police moved in suddenly and without warning and started hitting protesters with batons, pushing them off the monument and dragging them away."[16]

The operation lasted just twenty minutes. HRW reviewed video footage showing "riot police moving in on the protesters, striking them with batons, and kicking and hitting people who fell." There were also reports of tear gas and stun grenades. Some sixty to ninety-one people were injured, according to a later report by the International Advisory Panel.

The crackdown backfired. Later in the morning of November 30, the crowds in the Maidan more than doubled to between five hundred thousand and a million people. Protest organizers became more aggressive as well. On the afternoon of December 1, a group of fifty to sixty protesters, led by a well-known Ukrainian journalist, Tetiana Chornovol, broke into the offices of the Kiev city administrator, claiming it as their own headquarters. Later that afternoon, officers of the Ukraine's special security police—the Berkut—counterattacked. Hundreds of protesters and some fifty police officers were injured in the ensuing clashes.

The demonstrations, which had started peacefully, were de-volving into almost daily clashes between police and protesters. In the days that followed, police tactics became more aggressive. Yet the crowds only grew. On December 8, another protest drew hundreds of thousands of people.[17]

Ukrainian government officials and their Russian allies were growing impatient. On December 17, 2013, President Yanukov-ych traveled to Moscow to meet President Putin. Putin and Yanukovych—patron and supplicant—agreed on a joint action plan. In the days that followed, the crackdown grew worse. Within a week, the number of law enforcement officers, including police, Berkut, and so-called internal troops, had doubled to more than ten thousand. The first protester died after being severely beaten by unknown assailants. The journalist and protest leader Chornovol was also severely beaten and hospitalized. However, the demon-strators did not back down—they took their protests right to the doorstep of President Yanukovych's official residence.

In January, Ukraine's assembly passed what came to be known as the "Draconian Laws," which increased penalties for public demonstrations and other offenses, including wearing masks. More ominously, police began firing on protesters with rubber bullets and grapeshot. Protesters were beaten and kidnapped. Three more were killed in January. The number of police and law enforcement officers deployed in Kiev tripled to thirty thousand.[18]

In Washington, fears grew of a Tiananmen Square–style mas-sacre. And, in fact, internally, Russia was discussing just such a crackdown. Subsequent investigations revealed what appeared to be authentic Russian military plans calling for sending tanks into the square and deploying helicopters that would fast-rope spe-cial forces into the union building where the protesters had their headquarters.

"There was one note [in the Russian plans] that struck me as particularly Russian, which was 'Be sure to find [opposition politicians] Yatsenyuk, Klitschko, and Tyahnbok and kill them all,'" recalls Pyatt. "So that there's no way that this can reignite at some point in the future."

The Kremlin was making plans to massacre both the demonstrators and Ukraine's independent political leadership.

In the Kremlin, a siege mentality had set in. Putin saw Washington's hand in every move by the protesters and their leaders. From the beginning, Russia suspected that the United States was orchestrating the protests. They blamed the Obama administration generally, and Hillary Clinton specifically. This belief would later help fuel Putin's personal animus against the secretary of state, which would in turn fuel his motivations for interfering in the 2016 presidential election to her opponent's benefit.

In the first week of February, as the protests gained steam, the Kremlin would seem to have obtained proof of American meddling. On YouTube, a recording of a phone call between Assistant Secretary of State Victoria Nuland and Ambassador Pyatt appeared. It wasn't clear who had intercepted the call, though Russian intelligence was widely suspected.

On the recording, Nuland and Pyatt discussed their support involvement with the leaders of the Ukrainian opposition. At one point, for instance, they discuss discouraging Vitaly Klitschko, one of three main opposition leaders, from joining a new government.

Nuland says, "Good. I don't think [Klitschko] should go into the government. I don't think it's necessary, I don't think it's a good idea."

Pyatt responded, "Yeah. I guess . . . in terms of him not going into the government, just let him stay out and do his political homework and stuff. I'm just thinking in terms of sort of the process moving ahead we want to keep the moderate democrats together."

Later, Nuland crassly dismissed European involvement with a line that made headlines across the continent. She expressed relief that the United Nations was getting involved, before adding, "That would be great, I think, to help glue this thing and to have the UN help glue it and, you know, fuck the EU."[19]

In June 2017, Nuland would attempt to explain the comments to PBS's *Frontline*: "We had been trying to get the EU to be the midwife of the negotiations for weeks and weeks, and they were cautious about it. So, we had been planning another alternative, which was the UN. The phone call is about the ambassador saying to me: 'OK, we finally have a break. Yanukovych has made this offer of a couple of jobs to the opposition, but we need a moderator, a moderating force for these negotiations. Do you think the EU will play?' And I say: 'OK, eff the EU. We need to move on to the UN because we don't have any time, and the EU had three weeks to make this decision.'[20]

"So that's what that was about," Nuland told PBS. "It was not about a cosmic judgment about the EU. It was a tactical, urgent decision to try to get people off the street, get a peaceful government solution."

Regardless, the call's release was an embarrassment to the United States. More important, it would further fuel Putin's paranoia toward the United States in general, and Hillary Clinton in particular, a paranoia that would help spark an ambitious attack on the US presidential election beginning the following year.

On the Maidan, the protests would take a bloody turn. On February 18, at least eight protesters were killed and more than one thousand injured. Negotiations between President Yanukovych and protesters broke down the next day. The Security Service of

Ukraine ordered an "antiterrorist operation" to clear the Maidan once and for all.

A later report by the International Advisory Panel documented the alarming next steps: "A Berkut unit [of Ukraine's special police] . . . moved up Instytutska Street armed with sniper and Kalashnikov rifles and shooting, in particular, from the barricades situated near Khreshchatyk metro station.

"Between 8:20 am and 10:00 am [on February 20]," the report found, "49 persons were shot dead." Ukraine's Ministry of Health would report that "106 persons perished and died on the territory of Ukraine" between November 2013 and February 2014, with at least 78 deaths in and around the Maidan. Forty journalists had been beaten and severely injured. An uncertain number of protesters had disappeared. The actual number remains in dispute. Thirteen law enforcement officers were also killed in the clashes.[21]

The alarming death toll was too much even for Yanukovych. On February 21, he and opposition leaders signed an agreement restoring Ukraine's 2004 constitution granting more power to the parliament, forming a new coalition government, and scheduling an early election for a new president of Ukraine by December 2014. That night, Viktor Yanukovych fled Kiev.

In the days that followed, Yanukovych was a fallen president without a home. Mystery surrounded his location. Was he in Ukraine or Russia? Would he come back? The US embassy in Kiev—the entire US government—became a victim of the confusion as well. It had lost track of the president of Ukraine—lost track of the leader of one of the largest countries in Europe.

"We literally lost Yanukovych for a couple of days," Pyatt recalled. "He was—you know—Where's Waldo?"

Yanukovych first flew to Kharkiv in Eastern Ukraine, believing he could rally his supporters there. But he was disappointed to find little backing. Even in the east, word that he had ordered the shooting of peaceful demonstrators in Maidan made him persona non grata among the general public. He had also lost his closest confidants: Ukrainian oligarchs.

"The oligarchs around him fled. They said, this is the guy they had sold all their stock for and they realized that he was losing assets," Pyatt said.

This was the kiss of death for the Ukrainian president. He had lost his political and economic stock in Ukraine and, more important, in Russia. Vladimir Putin does not suffer fools. If anyone was going to take the blame for failing to suppress the demonstrations, it would not be the president of Russia.

Inside Ukraine, the days that followed the departure of their elected president were an almost mythical, if ultimately brief, taste of the Western-style democracy many of them had dreamed of.

"I told Ukrainians, that was your one weekend of happiness," said Ambassador Pyatt. "Because it was. It was amazing to see the mood in Kiev in those days, because on the one hand you could still smell smoke everywhere, but there were also candles and flowers. I've never seen so many flowers. It was grandmothers and widows and kids.

"It must have been what it was like after Gettysburg," he continued. "This incredibly traumatic, bloody, convulsive experience. But then everybody said there is a collective sense of coming together. We have to make this work."

However, Ukraine was suddenly a country without a government.

"I looked outside my house and the police watch post in front of my house was vacant."

Concerned about the safety of his staff, which included dozens of Americans and their families, Pyatt reached out to Petro

Poroshenko, the chocolate magnate and opposition leader, for assurances about their safety.

"I said, 'Hi, I just want to let you know we're not really sure who's in charge right now, but my foremost responsibility is the safety and security of American property and my personnel, and can you please get me somebody—whoever it is and whoever is in authority right now.'

" 'There's no police and the police have just disappeared!' "

As Yanukovych continued his long, meandering trip to Russia, his underlings attempted their own escapes. Security cameras at the airport in Kharkiv captured one Yanukovych crony running down the corridor toward a private plane with a hockey bag stuffed and heavy.

"We don't know if it was gold bars or whatever, but he was running through the airport so fast that he pulled down the whole magnetometer.

"For me, it captures the end of the Yanukovych cohort because it was not dignified," said Pyatt. "They fled, and they fled because they realized their whole house of cards was collapsing."

Average Ukrainians were now getting a window inside this gilded house of cards for the very first time. Like the US ambassador's residence, the security posts surrounding the presidential mansion were also suddenly empty, opening the grounds to hundreds of Ukrainian citizens and their smartphones.

"They walk in to discover all the llamas and the car collection and sort of his own Bedminster, all that crazy shit," said Pyatt.

Videos from inside the mansion went viral across Ukraine and soon around the world. For this reporter, the scenes brought back memories of Iraqis storming Saddam Hussein's palaces after the US invasion in 2003. But there was a difference here. Despite seeing the evidence of Yanukovych's looting of their national wealth, there was no looting or vandalism.

"It was really quite remarkable that amid all of that, there was no retribution," Pyatt said. "There was no fury."

Inside the Kremlin, the anger was palpable and paranoia at fever pitch. Russian president Vladimir Putin was in the midst of his prized Winter Olympics in Sochi, what had previously been a modest ski resort some three hundred miles down the coast from Crimea. He had spent some $51 billion on the Games, four times the original budget and nearly eight times the cost of the previous Winter Olympics in Vancouver. He hoped Sochi would serve as an international symbol of growing Russian wealth and might. Now, he suspected a CIA-orchestrated coup in Ukraine, timed precisely and intentionally to mar his Olympic games.

"I remember there was some reporting suggesting that he was angry because he believed this was part of the effort to humiliate him," recalled Pyatt.

Putin's perceived humiliation would prove to be powerful and lasting. The United States had its own political process approaching, the 2016 presidential election, with a leading candidate who was a particular target of Putin's anger and resentment. His more immediate target, however, was much closer to home.

"It was clear at that point that somebody pulled out a plan that existed already and said, 'We're going to invade. We're going to show those bastards,'" said Pyatt. "'We're going to invade Crimea.'"

———

The timeline of events in Ukraine in late February 2014 was remarkably swift, revealing Russia's execution of the Shadow War in stark clarity. This was a chain reaction of alarming speed and aggression, extending from the Maidan to Crimea to Eastern Ukraine and, soon, into the skies over Europe.

On February 22, Yanukovych fled Ukraine, leaving what the Kremlin viewed as a satellite state in the hands of a pro-Western government following protests Putin saw as a coup orchestrated by the United States.

On February 23, Russia held the closing ceremony for the Sochi Olympics, an exceedingly expensive event that Putin had intended to showcase a new, more powerful Russia to the world.

On February 25, the first units of what would come to be known as the "little green men" appeared in Crimea.

Once again, the US government and much of the US foreign policy establishment were caught off guard.

"The whole of our policy at that point was to help Ukraine re-stabilize itself, reconstitute its democracy, and get on with the process of regularizing," said Ambassador Pyatt. "They needed new elections. They needed to legitimize the new political order. They needed to heal after the violence of the Maidan.

"There was nobody—nobody—in the US government who predicted that the Russian response would be so expansive and militarized," said Pyatt. "There was a failure of imagination because we were mirror-imaging."

US intelligence agencies claimed to me at the time that their intelligence assessments had included the possibility of military action by Russia.

Shawn Turner, spokesman for Director of National Intelligence James Clapper, released a statement on March 5 saying the intelligence community had warned the administration a week earlier that Crimea was a "flashpoint for Russian-Ukraine military conflict."

Turner said that assessment included an analysis of Russian

military assets "staged for a potential deployment and those already in Ukraine that could be used for other purposes. It clearly stated that the Russian military was likely making preparations for contingency operations in the Crimea and noted that such operations could be executed with little additional warning," he added.[22]

John Scarlett, the former head of Britain's MI6, also disputes a failure by Western intelligence in Crimea, but for a different reason. He doubts that Vladimir Putin actually had a plan to send those "little green men" into Crimea. In his view, Putin was making it up.

"I've tended to see the decision to move in as probably quite last-minute," said Scarlett. "They were reacting to the sudden crisis in Ukraine itself."

In his view, Russia was caught off guard by the durability of the Maidan protests and then the swift collapse of President Yanukovych and the pro-Russian government that Moscow had nurtured for years.

"With hindsight, everybody thinks there is a plan. In practice, things are often day-by-day and more move-by-move," said Scarlett. "My impression is there was a lot of reactive decision making [by Russia] involved in the Crimea and Ukraine crisis."

The reactive decision making extended all the way to Washington, if there was decision making at all. Inside the State Department and the White House, officials were still deliberating how to define what Russia was doing, and whether Russia had in fact broken international law.

"There were lots of conversations with lawyers litigating over what is permissible under the Black Sea Fleet agreement and whether we call out the Russians for violating their treaty obligations," recalled Pyatt.

Events on the ground were rapidly outpacing deliberations inside the US government—a constant feature of the Shadow War. Yet many Russia experts inside the government continued to insist that Putin's occupation of Crimea was temporary.

"There were lots of Russian experts who said he'll never annex Crimea. He would never do that," said Pyatt. "That would be far too provocative. He'll send some troops in to destabilize and send a message but he's not going to go so far as to actually change the borders of the Russian Federation."

In fact, Vladimir Putin was preparing to change the borders of both Russia and Europe, shattering a bedrock principle of the post–World War II order: that European powers would not and could not alter the borders of sovereign states by force.

On March 18, less than a month after the first Russian forces poured into Crimea, Putin would make Russia's annexation of Crimea official in a bold and defiant speech at the Kremlin.

"In people's hearts and minds, Crimea has always been an inseparable part of Russia," Putin said. "This firm conviction is based on truth and justice and was passed from generation to generation, over time, under any circumstances."

He closed his address by declaring, "Today, in accordance with the people's will, I submit to the Federal Assembly a request to consider a constitutional law on the creation of two new constituent entities within the Russian Federation: the Republic of Crimea and the city of Sevastopol, and to ratify the treaty on admitting to the Russian Federation Crimea and Sevastopol, which is already ready for signing."

He added, addressing Duma members, "I stand assured of your support."[23]

"That was Putinism. It was the most naked manifestation of his revisionist agenda," said Pyatt. "And he did it in a *Dirty Harry* way. 'What are you going to do about it?' He was using military force to establish a political fait accompli and then challenging all of us to do something about it."

The United States and Europe did not appear to hear Putin's message. US and European officials expressed support for the protesters and for Ukraine's fledgling new government but had no clear plan to challenge Moscow's annexation of Crimea. Ambassador Pyatt watched this failure play out during a visit to the Maidan by Secretary Kerry in April.

"Literally, there was still ash everywhere. It smelled like smoke," Pyatt recalled. "We pull up in the big Cadillac. Kerry gets out and everybody starts clapping. America!"

Pyatt recalls hearing some elderly Russian women off to the side, asking one another, "Who is this guy?"

One answered, "I'm not sure who he is but he's American. He must be good."

The scene was not one Pyatt or the State Department expected or desired. The Kremlin was already convinced the popular protests in the Maidan were the work of the US government. Russia's foreign minister Sergei Lavrov was publicly referring to the protests as a "coup d'état," laying the blame on Ukrainian "fascists" backed by Americans.

As he was returning to his limousine, Secretary Kerry asked one of the protesters through the embassy interpreter why he endured the cold and the violence to remain there in the Maidan.

Pyatt remembers the protester's answer vividly. "He said, 'Because I wanted to live in a normal country,'" said Pyatt.

Kerry's appearance resonated across Kiev all the way to Moscow. However, the US policy response remained cautious and halting.

"Kerry was still talking in terms of 'Russia must not overstep,'" said Pyatt. "And it was while they were already running the place."

Inside the Obama administration, discussions focused on providing Moscow with a diplomatic "off-ramp" to defuse the crisis and eventually exit Crimea in a face-saving way.

"It goes to the fundamental challenge that we face with Russia today which is that we are dealing with a government that does not believe in win-win," said Pyatt.

Along with its deployment of "little green men" in Ukraine—that is, Russian regulars with only their name tags and unit patches removed—Russia executed another tactic of the Shadow War, a now-familiar Orwellian assault on the truth.

"You sow confusion and you sow doubt," said Pyatt. "And that in turn reflects what I think is the most pernicious aspect of this new Russian strategy of information warfare: the recognition that you can achieve political and diplomatic effects by manipulating information.

"The Russian objective was not to win the argument," Pyatt emphasized. "It was to win a war."

As the United States and Europe dithered over Crimea, Russia executed its next aggressive move. To the northeast, in the portion of Eastern Ukraine bordering Russia, Moscow would follow a similar pattern as it had pursued two months earlier in Crimea. In April, as Russia was formalizing its annexation of Crimea, armed clashes broke out in the Donbass region between Ukrainian forces and pro-Russian separatists. There was no question as to who was backing the separatists.

Then, in August, claiming ethnic Russians were again under assault, Russia openly moved its own forces into Ukraine. Six months after invading Crimea, Russia had begun a second invasion of sovereign Ukrainian territory—and its second violation of the borders of a sovereign European nation.

Once again, US officials still debated just how deeply Russian military and Russian intelligence services were involved in the military maneuvers in Eastern Ukraine.

"You had the same dynamic where we were trying to diagnose what was happening on the ground," recalls Ambassador Pyatt.

Again, US intelligence agencies came under fire for failing to foresee Russian actions. Ukrainian officials and the Ukrainian public, however, began to fear something far worse: Russia's annexation of all of Ukraine, repeating the Soviet Union's absorption of their nation following World War II.

"In that April and May period, there were days when I had my local employees at the embassy coming to me with images of the last helicopter off the roof in Saigon, saying, 'Mr. Ambassador, when you Americans leave, what happens to us?'"

Like Estonians who feared that the 2007 cyber assault presaged invasion, Ukrainians feared Russian-backed military actions in Crimea and Eastern Ukraine were just the first steps of their own takeover.

"They were very genuine concerns," Pyatt said. "Ukrainians have living memories of what it feels like to be absorbed."

On the approach to Kiev, Ukrainian forces began setting up tank traps on the highways. Fortified positions went up on the eastern side of the Negro River as part of a final defense of the capital. Ukraine's then–prime minister, Arseniy Yatsenyuk, took Ambassador Pyatt aside for a dire warning.

"I remember a late-evening conversation with Prime Minister Yatsenyuk. He said to me, 'You know you guys need to understand that if we lose, and if Putin prevails, and if these forces come all the way to Kiev, I will be dead. Everyone I care for and my family will be killed or thrown in jail.

" 'We have lived through this. We saw what happened with the Nazis and then what happened when the Nazis were driven out and Stalin came back,' " Yatsenyuk told Pyatt.

As the Ukrainian prime minister contemplated his own murder at the hands of Russian invaders, US and Western officials were still debating how to react. The divisions within Europe were the widest. Great Britain and France advocated for a robust response. Germany, which maintained the closest business and diplomatic ties to Moscow, pushed for patience.

Some European diplomats believed only American leadership could unite the West against Russian aggression.

Pyatt recalls a desperate meeting with a Polish diplomat and longtime friend.

"He said, 'You know, Geoff, you have to keep pushing, because if America doesn't lead on this, Europe isn't going to. Europe isn't going to come together,' " recalled Pyatt.

US leadership would not materialize that summer, however. The United States and its European partners remained largely in discussion mode as Russian forces solidified their grip on Crimea and made further territorial gains in Eastern Ukraine.

———

It would take a passenger jet falling out of the sky over Europe to shake US and European leaders out of their stupor. The crime was simply too appalling and the evidence of Russian responsibility simply too clear.

Alexander Hug and his team from the OSCE found themselves on the front line of establishing the facts from the moment they arrived at the crash scene on July 18, 2014. They would return every day for three months, gradually cajoling and negotiating their way with Russia's "little green men" to make their way farther into a crash site that extended for miles. They were not crash investigators. No one on his team had ever visited a crash site. But they were the only formal observers able to determine the facts of what increasingly appeared to be an appalling crime.

A plane crash provides its own unique glimpse of hell. The vagaries of physics and luck mean that some bodies are torn apart, or burned beyond recognition, while others fall to the ground seemingly untouched.

"I just tried to imagine what actually happened on that plane," he said, the images of the scene still vivid in his mind. "If you see the bodies and the debris and the burnt areas, it made me think. What were their last moments? What were they doing? Were they sleeping? Did they feel anything? Did they actually feel the crash?"

Walking the main debris field, Hug came across a row of seats, lying perfectly upright with a handful of passengers still buckled in.

"We couldn't see any visible injuries on some of these bodies. They were intact," he said. "They had no painful expression on their faces. Maybe my mental image is blurred with emotions now, but that was at least what I felt at the time."

Flight MH17, packed with families heading to vacations in Asia, carried an unusually high proportion of children. Of the 298 passengers and crew on board, 80 were kids. As a father of three himself, Hug found the scene jarring.

"The toughest part were the kids. They always look innocent," he said. "I often recall these moments. They come up as flashbacks. Gruesome stuff and it always comes together: on the one side, you remember this image and on the other side, you see your own kids."

Within the carnage, small gestures took on special meaning.

Hug wore a tie every moment he was on-site. The victims deserved respect—and he was intent on demonstrating it any way he could.

"I felt some responsibility to bring some dignity to the crash site, some normality, because we knew the whole world—and all their relatives—were watching," he said. "We felt it was part of our job."

Local residents took particular interest in collecting children's possessions. A pile of stuffed animals grew to several feet tall. There were schoolbooks, backpacks, and sippy cups. One journalist brought Hug a handful of passports he had collected.

"He had a stack of passports that he had found, and he wanted to come into one of our cars to talk to me," Hug remembers. "He handed them over to me and said, 'Make sure they get back to where they need to go.' Then he broke into tears."

By October 13, nearly three months after the jet went down, they had facilitated the recovery of most of the victims and their personal belongings, handing them over to Dutch authorities. The OSCE and other teams would still be recovering human remains and debris, however, well into November.

———

In September 2016, investigators from the Joint Investigation Team (JIT) for MH17, composed of representatives from countries that had lost the most citizens in the disaster, including Australia, Belgium, Malaysia, and the Netherlands, plus Ukraine, delivered a report on the crash to relatives of the victims. The JIT, whose focus is on prosecuting the perpetrators of the attack, declared that "irrefutable evidence" indicated a Russian-supplied missile fired from territory controlled by pro-Russian separatists had brought down the jet.[24]

Though the United States had reached a conclusion within

twenty-four hours that it was a Russian missile that brought down MH17, America's European allies were eager for an independent inquiry to establish the facts. The Netherlands, which had suffered the biggest human loss, took the lead.

Like the analysts at the Defense Intelligence Agency in the hours immediately following the attack, Dutch investigators established the explosion that brought the plane down took place outside the fuselage.

They then established there was no other aircraft in the area that could have fired the fatal shot. As the Joint Investigation Team report concluded, "sufficient radar data" from both Ukraine and Russia showed that "at the time of the crash, no other airplanes were in the vicinity that could have shot down flight MH17."

Dutch investigators intended to meet a standard even Russia could not contest. Like detectives collecting evidence of a murder, they had to establish a hard link between the weapon and the aircraft and its victims. They were undertaking a high-stakes game of Clue, in front of an international audience.

A BUK missile, like most surface-to-air missiles, does not strike its target head-on, but rather explodes in front of it, the blast and supersonic spray of shrapnel tearing the aircraft apart in the sky and bringing it down.

The Joint Investigation Team found the most definitive evidence inside some of the victims' bodies. Its report stated bluntly, "During the autopsy of the bodies of the cockpit crew, several fragments were found that belonged to the warhead of a 9M38 series BUK missile."

The JIT findings continued, "One of these fragments showed traces of a cockpit glass on the surface, which was the same unique type of glass that is used for a Boeing 777."

And the JIT found further evidence in the cockpit itself. "In the frame of one of the cockpit windows a metal piece was found which was identified as part of a 9M38 series BUK missile. This piece was located in a twisted position in the frame, making it clear that it was shot into the window frame with great force," read the report.

Citing witness statements, photographs, videos, and intercepted conversations, Dutch investigators accomplished something no other investigating body had yet managed: tracing the entire path of the BUK missile system from Russia into Ukraine to its eventual firing position near the town of Pervomaiskyi and, after the fatal shot, back again into Russian territory.

"During the night [of July 14]," the report found, "the convoy crossed the border into the territory of the Russian federation."[25]

In the end, the deadly missile launcher and its crew spent less than twenty-four hours inside Ukrainian territory.

———————

M H17 did what Russia's annexation of Crimea and invasion of Eastern Ukraine had not: unified European leaders that Russian military aggression inside Ukraine—and more broadly, inside Europe—had gone too far, and would go further if not met by imposing sufficient costs on Russia. Establishing what constituted sufficient cost to deter would become a consistent challenge for Western leaders.

Where the United States and the West did fail was in the pace and the strength of their response. Western allies were divided, first on Russia's responsibility, then on its intentions, and then on the best way to deter further aggression. US leaders were slow to recognize the facts on the ground, divided themselves as to how far the Kremlin was willing to go. When the United States and its allies were finally on the same page on Russia's culpability, they were not on penalty and deterrence. In fact, they are still debating their response today, with the US president leading the push for a more conciliatory approach to Russia.

Ambassador Pyatt expresses a palpable sense of frustration and regret with his own government's reaction both in 2014 and today.

"I will never know what would have happened if the sanctions had been implemented with greater alacrity—and if we had been able to get the Europeans to take it seriously and get the sanctions," Pyatt told me.

One element that was lacking from the beginning was clear leadership from the United States.

"When I arrived in Kiev," said Pyatt, "my instructions were: Europe is in the lead."

Pyatt frequently recalls the Ukrainian protester in the Maidan who came face-to-face with Secretary Kerry.

"It's the guy with Secretary Kerry saying I want to live in a normal country under the rule of law, courts, freedom of speech, and freedom of the media," he said. "But to the extent Putin chooses to play a hard-power card, it has to be the United States in the lead."

By the end of 2018, Russia's invasion had caused more than ten thousand deaths in Ukraine, from the battle for the "republics" in the East, to the annexation of Crimea, to the crackdown on peaceful civilian protesters in the Maidan, to the 298 passengers and crew of MH17, shot out of the sky over Europe.

At the G7 summit in 2014, President Obama had dismissed Russia as a "regional power," saying that its territorial ambitions "belonged in the nineteenth century."

"The fact that Russia felt it had to go in militarily and lay bare these violations of international law indicates less influence, not more," Obama said.

His comments in 2014 echoed his dismissal of his Republican

opponent Mitt Romney's foreign policy priorities in an October 2012 presidential debate: "When you were asked what was the biggest geopolitical threat facing America, you said Russia, not Al Qaeda. You said Russia and the 1980s are now calling to ask for their foreign policy back because the Cold War's been over for twenty years."

Romney's answer now looks prescient. "Russia indicated it is a geopolitical foe," he said. "I'm not going to wear rose-colored glasses when it comes to Russia or Mr. Putin."

But Russia's nineteenth-century tactics had, for now, bested the West's twenty-first-century politics and diplomacy.

Some Russia analysts and policy makers in the United States and the West argue that the West goaded Russia into its land grab in Ukraine by overextending the West's influence into Russia's "near abroad." The implication is that Russia's invasion of Ukraine was a reaction, in effect, to the Western overreach.

When I asked him about this argument, Ambassador Pyatt told me he has one word for it: "bullshit."

"It goes back to my conversation with [EU Commissioner for Enlargement] Stefan Fule in Yalta in 2013. It was not the United States or the Europeans who were expanding into Russia's strategic space. It was the citizens themselves who needed to decide.

"The idea that somehow the United States is to blame for this misses the idea—misses the fact—that these individuals—these European citizens—have free will. There was nobody from the US government who was ever pushing them."

Former MI6 head John Scarlett sees a subtler failure, that is, a failure to recognize the depth of Russia's alarm at Ukraine's drift toward the West—and its willingness to prevent that drift with alarming aggression.

"I'm going back to the point which is understanding the particular role that Ukraine plays in Russian decision making and in Russian priorities and values," Scarlett explained. "We've really got to get our heads around that. It's a unique situation. We've just got to be so careful and responsible about how we speak about it and what points we raise and promises we make."

For Scarlett, the United States' and Europe's handling of Ukraine's potential admission to the European Union and the Maidan protests that followed once again demonstrated the West's failure to understand the mind of its adversary. In reality, the West and Russia had diametrically opposed views of Ukraine's rightful future.

"In our minds, it's become a separate independent country and of course it is—and we strongly defend that—but that I doubt very much in most minds in Moscow, they see it like that and I'm sure they don't," Scarlett said. "The idea it will develop in a separate way from Russia goes to the very heart of all the issues—the emotions—that exist after the collapse of a superpower."

Scarlett does not blame the United States and Europe for Russia's annexation of Crimea and invasion of Eastern Ukraine. But he does believe the West bears responsibility for misreading Putin and therefore misreading his interests and how far he was willing to go to defend them. That is—by definition—a failure of intelligence.

"You could argue not that we did the wrong thing, but we didn't quite understand the significance—the implications—of the things we did and said," said Scarlett. "The lesson there is—and part of the solution is—we just have to be better at understanding the mind of the other side.

"We have to ask ourselves the question whether we really understood that adequately," he said.

Nearly a year after the loss of MH17, in April 2015, Hug would have a surprise, repeat encounter with the armed commander who had first confronted him at the crash scene.

"I didn't recognize him because he wasn't drunk, and he looked much fitter," Hug recalled. "But he recognized me, so he came up to me."

The chance meeting took place in the small village of Shyrokyne, in southern Ukraine on the shores of the Sea of Azov. This was a new front in the continuing war in Ukraine—the site of a six-month standoff between Ukrainian and Russian-backed forces.

"It was a tense interaction because that village had been shelled and it's destroyed," said Hug, but his old friend attempted to make light of the situation.

"He said, 'Alexander, what are you doing here? There is no Boeing here,'" Hug told me. "And then I realized who the guy was."

Hug was astonished to see him alive. After MH17, he and the OSCE had lodged complaints with his higher-ups about him. And, in a rare concession, they had removed him.

"I thought he would not have made it out. He disappeared from the scene," said Hug. "We thought he had been sent off. We were even told he had been sent to dig trenches."

And yet here he was nearly a year later on another front of the Shadow War.

"He had replaced his camouflage hat with a red beret, but he still had no insignia on, and was still clearly doing the same job that he had years before," Hug said.

Like Russia, the fiery officer was still in Ukraine with no intention of leaving.

LESSONS

Russia's annexation of Crimea and invasion of Eastern Ukraine in 2014 presented the United States and the West with more stark lessons. First, it demonstrated that the Kremlin had both the intent and the ability to redraw the borders of Europe by force. And it was willing to do so right on NATO's doorstep. Russia's aggression was a qualitative step beyond its invasion of Georgia in 2008 in that Ukraine lies within the boundaries of Europe and right on the border of four US treaty allies: NATO members Romania, Hungary, Slovakia, and Poland. At the same time, Moscow was making clear that it would not let NATO or the European Union expand into territory it considers its "near abroad."

Second, the United States and the West missed or ignored repeated warnings of Russia's intentions, including explicit threats to exert military influence over Ukraine by President Putin and other officials in the years and months before "little green men" showed up on the streets of Crimea. The West's mistaken view that Russia was willing to play by the West's rules persisted even as tanks rolled across the border into a sovereign European nation. And today, years after the invasion, Crimea and large portions of Eastern Ukraine remain under Russian control. These "facts on the ground" make clear that the US policy of imposing economic sanctions on Russian leaders and entities has not raised the costs on Russia sufficiently to deter or change its behavior. Today's new reality in Ukraine raises an even more disturbing question central to the Shadow War: if Russia was able to seize territory inside Europe without a military response from the United States and the West, would it be willing and able to do the same to a NATO ally such as Estonia? It's a question and a threat that remains unresolved.

CHAPTER 5

Unsinkable Aircraft Carriers

(CHINA)

On the tarmac of Clark Air Base in the Philippines, the P-8A Poseidon was at first hard to distinguish from the commercial 737s parked nearby. Built on a Boeing 737-800 frame, the P-8 looks less like a warplane than a charter jet. Up close, though, its military capabilities become clearer. Peeking out of the fuselage are arrays of antennas, radar domes, and camera wells. Cut into its belly are hatches for dropping sounding buoys and torpedoes. Under the wings are wells, empty on this trip, for Harpoon missiles. Inside I had the feeling of entering a CIA listening station in the sky. The Poseidon is packed with advanced intelligence-gathering equipment; a dozen crew members sat at screens lined up and down the center of the fuselage. Despite appearances, this is a weapon of modern warfare—the Navy's most advanced surveillance and sub-hunting aircraft.

My CNN colleagues Jennifer Rizzo, Charles Miller, and I were the first journalists allowed on board the P-8 for an operational mission. It was May 2015 and tensions between the United States and China were rising over what the United States saw as an unprecedented acceleration of land reclamation activities in the South China Sea. On board, I sat in on the flight crew's "delta briefing," the final meeting before takeoff. The aircraft's commander, Lieutenant Commander Matt Simpson, laid out the mission plan. The P-8 would leave Clark and fly some 460 miles west across the South China Sea to the airspace over three reefs—Fiery Cross, Subi, and Mischief. The reefs until recently had been uninhabitable piles of rocks, barely peeking above the surface at low tide. Since 2012, however, China had rapidly transformed them into man-made islands, which the United States feared would become permanent military installations—"unsinkable aircraft carriers"— some six hundred miles from the Chinese coastline and right in the middle of waters claimed by some half a dozen neighboring countries, including the Philippines, an ally the United States is obligated to defend against any and all military aggression.

Within weeks of entering service in November 2013, the P-8 deployed to Asia. Its principal mission: keeping a watchful eye on Beijing as it expands military operations in the region. Clark Air Base in the Philippines, with its proximity to the South China Sea, would be an important staging ground. The US military had had a continuous presence at Clark since the turn of the last century, flying missions during World War II, the Vietnam War, and, later, the Cold War. But in the early 1990s, with the end of the Cold War and increasing domestic opposition to the US military presence, Washington and Manila signed an agreement to transfer Clark to the Philippine military. The last US combat aircraft were transferred out in 1990.[1] Twenty-five years later, however, US military aircraft were back and their flight crews were busy.

Takeoff from Clark was swift on the Poseidon. There was no slow roll down the runway, just a rapid revving of the engines to full power and, within seconds, we were in the air—the turquoise blue of the South China Sea coming into view quickly. As we made our way through clear skies, the P-8 crew, composed of more than a dozen naval aviators, invited me into the cockpit to watch and listen. The cockpits of modern military aircraft are marvels of technology. I sat in the jump seat just behind the pilot and copilot, watching the pilot track the jet's course via green crosshairs superimposed on the windshield. It made flying look as easy as a video game: keep the X in the middle of the circle, and all was fine. Watching the pilot's hands on the yoke belied that impression, as she made constant micro-adjustments to keep the jet flying level. The crew would soon be entering contested airspace.

Forty-five minutes into the flight, the first target came into view: Subi Reef. In its natural form, Subi was a thin, oblong circle of sand and rocks shaped like a giant carabiner, surrounding a deep lagoon—a quiet, useless plot of land and occasional refuge for fishing boats. Now it was a hive of activity. More than two dozen Chinese dredgers filled the lagoon, pumping sand from the ocean floor in huge plumes onto the surface, slowly and gradually enlarging and fortifying the reef. China was building an island from scratch.

The scale and pace of the dredgers' work was mesmerizing. In the span of two years, China had expanded Subi and its close neighbors Fiery Cross and Mischief Reefs by two thousand acres— the equivalent of more than 1,500 football fields. It was an engineering marvel in waters as deep as three hundred feet.

"We see this every day," Captain Mike Parker told me with a smile. Captain Parker commanded the VP-45 maritime patrol squadron, the "pelicans," as they're known, based out of Jacksonville, Florida, and then deployed to Atsugi, Japan, with six new P-8s.

"I think they work weekends on this," he said.

The crew trained the P-8's high-resolution cameras onto the scene below to gauge progress up close. The dredgers' pumps sucked sand off the bottom of the sea and shot it like giant fire hoses onto the surface. Their work was brutal in its efficiency: create new land on the surface, while digging deep harbors below.

As we left the airspace over Subi Reef, we soon approached Fiery Cross Reef, just a few minutes' flying time away. Here China had made the most extensive progress. With land reclamation complete, China was outfitting a full-fledged air and naval station.

On the P-8's high-resolution video cameras, we could identify early-warning radar installations, military barracks, a control tower, and aircraft hangars hardened against aerial bombardment. The airstrip was long enough to accommodate every Chinese fighter and bomber. Chinese warships swarmed around the new islands, forming a protective perimeter.

"There's obviously a lot of surface traffic down there," Lieutenant Commander Matt Newman told me from the copilot's seat. "Chinese warships, Chinese coast guard ships. They have air search radars, so there's a pretty good bet they're tracking us."

Chinese dredgers continued to hollow out a deepwater harbor. The "unsinkable aircraft carrier" was nearly complete.

China had made repeated promises not to militarize these man-made islands. However, even from 15,000 feet up, those promises appeared empty. And even as construction continued, like the unfinished Death Star in *Return of the Jedi*, the islands were already carrying out military responsibilities—among them, warning away foreign military ships and aircraft.

In fact, the more China built, US commanders told CNN, the more frequently and aggressively the Chinese navy warned US military aircraft away. The P-8 crew was trained for these "challenges," and prepared to respond if necessary. Inside the cockpit, the radio soon crackled with a voice in Chinese-accented English: "This is the Chinese navy. This is the Chinese navy. . . . Please leave immediately to avoid misunderstanding."

"We were just challenged, and the challenge came from the Chinese navy, and I'm highly confident it came from ashore, this facility here," Parker said as he pointed to an early-warning radar station on Fiery Cross.

On both sides, the communications began calmly and formally. The Chinese navy operator declared this to be Chinese airspace and warned the US aircraft away. The US pilot listened and then read from a prepared script explaining that this was US aircraft operating in international airspace over international waters.

Over the next half hour, the US Navy flight crew and Chinese navy parried each other over the radio eight separate times. With each instance, I sensed frustration building from the Chinese radio operator. He came back on the line one last time, shouting with exasperation: "This is the Chinese navy. . . . You go!"

While the US Navy is trained for these interactions, civilian flight crews are not. The airspace over the South China Sea is busy with commercial aircraft connecting Asian cities to each other and to Europe and the West. Just after we heard the first of eight Chinese navy warnings to the P-8, the pilot of a Delta Air Lines flight spoke up on the same frequency, quickly identifying his aircraft as a commercial passenger jet. These can be nerve-racking interactions for civilian pilots.

On board the P-8, the mood was calm and confident. But the more China built, the US sailors told me, the more aggressively the Chinese navy challenged them. The crews were bracing themselves for a more dangerous kind of challenge. Once China deployed aircraft to the islands, US crews would face aerial intercepts. In 2001, in one such intercept near China's Hainan Island, a Chinese jet collided with a US EP-3 surveillance jet, the P-8's predecessor. The collision led to the fatal crash of the Chinese jet and severe damage to the EP-3, which managed a safe but difficult landing on Chinese territory. The deadly collision and subsequent detention of the US crew sparked an extremely tense standoff between the United States and China, which neither side hopes to repeat.

The P-8 flight crews know they are in the midst of a potential conflict. Chinese protests were becoming more aggressive. US surveillance flights and naval patrols were becoming more frequent. And the two countries' positions appear irreconcilable. China views these islands as sovereign territory and describes its commitment as "unshakable." The United States views the waters and airspace as international. Aboard the P-8, it was difficult to see how those differences will be resolved.

––––––––

From the air, the South China Sea appears peaceful and quiet— a calm, turquoise blue expanse seemingly more fit for island vacations than war. In fact, these are some of the most heavily trafficked and highly valued waters in the world. Ships carry some 60 percent of the world's trade through here. The waters cover some of the most fertile fishing grounds in Asia. Underneath the seafloor are rich reserves of oil and gas, still undeveloped. As a result, what had been uninhabitable reefs and shoals are suddenly valuable real estate.

Today the South China Sea is the subject of numerous competing territorial claims, involving China, Taiwan, the Philippines, Vietnam, Malaysia, and Brunei. Indonesia, while it has long insisted that it is not a claimant state, also recently renamed waters in its exclusive economic zone last year in an attempt to contest

China's own claims. To establish their sovereignty claims, each country reaches back deep into history, referencing ancient maps and the habits of fishermen dating back generations.

China's territorial claims in the region are the most expansive. Beijing relies on what's known as the "nine-dash line," an arbitrary demarcation line on a map drawn by what was then the Republic of China in 1947, to lay claim to virtually all of the South China Sea. The line extends from Hainan Island in the northwest to Taiwan in the northeast all the way down to the Spratly Islands in the southeast and extending up and down the coastline of the Philippines. China's man-made islands are the most concrete expression of those claims.

China first produced the map containing the nine-dash line to the United Nations Commission on the Limits of the Continental Shelf in response to what it saw as egregious claims on the part of Vietnam and Malaysia. Its roots, however, trace back to the original map produced by the Republic of China, the successor of which is now the ruling government in Taiwan. Bill Hayton notes that this claim was "the first ever made by any Chinese government in the Spratlys" with the potential to be recognized under international law.[2] The newly established People's Republic of China (PRC) under Mao Zedong would then take additional steps to assert sovereignty over more territory through the 1958 "Declaration on the Territorial Sea," which included claims to the Paracel Islands, Macclesfield Bank, and the Spratlys, as well as the new renegade province of Taiwan.[3] In the decades that followed, China would continue to assert sovereignty over much of what is now included in the nine-dash line map, formalizing its claims through domestic legislation in 1992 and engaging in a number of skirmishes with the Philippines and Vietnam.

America's own coastline is, of course, thousands of miles away, but the United States seeks to maintain the status quo, which holds that these are international waters, which means free passage for all international cargo ships and, crucially, the US Navy. Officially, the United States takes no position on each of the competing territorial claims. US policy through several administrations has been to support the resolution of those claims through international law and multilateral negotiation, rather than bilateral negotiation between China and its much smaller and much less powerful competitors or through unilateral action, including the construction of entirely new territory. Southeast Asian nations, fearful of being bullied or overpowered by Beijing, have generally welcomed America's role in defending freedom of navigation in these waters. At the same time, Beijing has leveraged its close political and economic ties with Laos, Cambodia, and Myanmar to prevent any meaningful opposition to its activity under the framework of the Association of Southeast Asian Nations (ASEAN).

While China's emphasis on its sovereignty over waters in the South China Sea has remained consistent since the PRC's founding, the creation of entirely new territory in the South China Sea is a new phenomenon—one part of a broader military push that some fear is intended to challenge overall US preeminence in the region. Beijing is sailing its first aircraft carrier; testing nuclear missiles with multiple warheads; deploying a vast network of missiles to destroy and deny access to US warships; and, now, building military bases far from its shores.

"I'm scratching my head like everyone else as to what's the [Chinese] endgame here," Captain Mike Parker told me.

Whatever its endgame, China's short-term calculation is transparent: the United States is not willing to go to war over these islands, so China is free to proceed. The US response has been to sail and fly through the area to demonstrate these waters and this airspace remain international and vessels should enjoy freedom of navigation as provided for in international law. These so-called freedom of navigation operations, or FONOPs, continue to contest Beijing's activity in the region. But the islands haven't moved, and in fact they are being further fortified militarily.

The Obama administration was alarmed by China's swift progress. Inviting a CNN reporter and his camera crew on board a P-8 mission was intended to alert the world to China's progress and to put Beijing on notice in a very public way. Our story, filed from the Philippines virtually the moment we landed on May 26, sparked headlines across the region and in the West. China's reaction, delivered by senior officials in the Foreign Ministry and the People's Liberation Army, was unusually stern.

"Why did this story suddenly pop up in the past few weeks? Has the South China Sea shrunk?" asked Senior Colonel Yang Yujun. "Some people have been intentionally and repeatedly hyping this topic. Their purpose is to smear the Chinese military and dramatize regional tensions. And I'm not ruling it out that this is being done to find an excuse for a certain country to take actions in the future."[4]

No one doubted that "a certain country" was the United States. China sought to turn the tables on the United States, claiming its own handling of the US Navy's "close-in surveillance" had been "necessary, legal, and professional." Colonel Yang was correct that such surveillance flights had been taking place for years, but this was the most public confrontation in the skies between the United States and China since the deadly high-altitude collision over Hainan in 2001.

Several months later, US and Chinese leaders would have an

even more public confrontation over the South China Sea. In September 2015, on President Xi Jinping's first official visit to the United States as Chinese leader, President Obama confronted his Chinese counterpart in public with US concerns.

With Xi standing beside him in the Rose Garden, President Obama said, "I conveyed to President Xi our significant concerns over land reclamation, construction, and the militarization of disputed areas, which makes it harder for countries in the region to resolve disagreements peacefully."

In a moment of strained diplomacy, President Xi did not back down, restating China's historical claims to the islands as fact.

"Islands in the South China Sea since ancient times are Chinese territory," President Xi said. "We have the right to uphold our own territorial sovereignty and lawful legitimate maritime rights and interests."

However, Xi made what the United States perceived as a promise as well, vowing that China would not transform the islands into military outposts.

"Relevant construction activity that China is undertaking in the Nansha Islands [the name China uses for the Spratly Islands] does not target or impact any country, and there is no intention to militarize," Xi said.[5]

While the Chinese word for intention (*yitu*) may fall short of an overriding promise to refrain from militarization, a rare display of such a commitment by China's leader resulted in the United States understanding it as such. Though the pledge contradicted the construction activity witnessed by the P-8 crews patrolling the South China Sea at the time, the White House took the Chinese president at his word.

Three years later, President Xi's "promise" looked not only empty but cynical. By 2018 Admiral Philip S. Davidson, soon-to-be commander of US Pacific Command, testified before the Senate Armed Services Committee that China had deployed sufficient military resources on the man-made islands to significantly challenge US military operations in the region.

"In the South China Sea, the PLA [People's Liberation Army] has constructed a variety of radar, electronic attack, and defense capabilities on the disputed Spratly Islands, to include: Cuarteron Reef, Fiery Cross Reef, Gaven Reef, Hughes Reef, Johnson Reef, Mischief Reef, and Subi Reef," Admiral Davidson wrote to the committee. "These facilities significantly expand the real-time domain awareness, ISR [intelligence, surveillance, and reconnaissance], and jamming capabilities of the PLA over a large portion of the South China Sea, presenting a substantial challenge to U.S. military operations in this region."

The bases, Admiral Davidson testified, were complete. The only thing they lacked were deployed forces.

"Once occupied," he warned further, "China will be able to extend its influence thousands of miles to the south and project power deep into Oceania. The PLA will be able to use these bases to challenge the U.S. presence in the region, and any forces deployed to the islands would easily overwhelm the military forces of any other South China Sea–claimants."

He included in his testimony this alarming assessment: "In short, China is now capable of controlling the South China Sea in all scenarios short of war with the United States."[6]

"All scenarios short of war." It was a perfect encapsulation of the Shadow War. How did China manage such a strategic victory is so short a time? The achievement was remarkable in both engineering and military terms. In five years, it had manufactured new territory in the midst of highly disputed waters and equipped them with advanced military capabilities—all hundreds of miles from the mainland. And it had done so with the United States and its neighbors imposing virtually no diplomatic or economic costs on Beijing. The answer to how China managed this coup involves a series of warning signs missed, and aggressive moves by China unchecked by the United States and its allies.

Andrew Erickson, professor of strategy at the US Naval War College, educates senior US Navy commanders on Chinese military strategy today and through history. Like Admiral Davidson, he believes that—with its man-made islands—China has achieved its goal in the region of creating "military-formidable bases capable of threatening foreign activities throughout the South China Sea."

In explaining how China managed such a remarkable territorial coup, Dr. Erickson points to a mistake similar to America's Shadow War with Russia: a gross US misconception of China's intentions and a stubborn unwillingness to recognize that misconception even when the evidence is as obvious as "unsinkable aircraft carriers" rising in the South China Sea.

"In a heartfelt effort to reassure China strategically and enlist its collaboration in noble global causes, US policy makers over the past decade have instead unfortunately emboldened China by looking weak and accommodating aggressive behavior," Erickson told me. "By failing to impose significant costs for harmful Chinese behavior at sea, the Obama administration inadvertently encouraged Xi Jinping to continue, and even increase, maritime malfeasance."

China's land grab in the South China Sea clearly demonstrates its "winning without fighting" strategy, that is, a perfect example of Beijing's own approach to waging—and winning—the Shadow War. Erickson notes the deep historical roots of the strategy, arguing that China's approach to the South China Sea epitomizes a strategy first articulated by the Chinese military strategist Sun Zi (or Sun Tzu) in the fifth century BC: "In war, the way is to avoid what is strong, and strike at what is weak."

"This applies not only to conventional major combat operations, but also to Beijing's approach to coercion short of warfare," says Erickson.

"Xi Jinping does not desire an all-out war with the US," explains Erickson, "but rather prefers to keep 'winning without fighting' in peacetime, or what the 2017 US National Security Strategy terms continuous competition—neither fully 'at peace' nor 'at war.'"

Like Russia, China is perfectly willing to flout international law and use military might to achieve its strategic interests with activities just below the threshold of war. China, however, is finding ways to achieve those interests in ways perhaps subtler than Russia. Unlike Russia in Ukraine, for instance, the Chinese military has barely fired a shot to achieve its objectives and acquire new territory in the South China Sea.[7]

China began pursuing this strategy in the South China Sea years before 2015, over another uninhabitable parcel of land just over three hundred miles to the northwest of the Spratlys, and right under the nose of the United States.

Like many land features in these waters, Scarborough Shoal bears almost as many names as countries who claim it. Scarborough, as it is labeled on Western maps, gets its name from the East India Company ship *The Scarborough*, which grounded on the rocks there in 1784. China refers to the same shoal as Huangyan Dao or Minzhu Jiao ("Democracy Reef"). The Philippines, another claimant, calls it Panatag Shoal, based on the Tagalog word for "threat" or "danger." Portuguese maps still use the name Bajo de Masinloc. The different names have meanings beyond their historical and linguistic origins. They are part of a broader effort by multiple countries to brand the Scarborough Shoal—like Fiery Cross, Mischief, and Subi Reefs—as long-held and legally defensible territorial possessions.[8]

The names are symbolic. In an apparent attempt to demonstrate that possession is 90 percent of the law, China began a bold effort beginning in 2012 to take control of the Scarborough Shoal, one fishing boat at a time.

Like many reefs and other land features in the South China Sea, Scarborough Shoal had no permanent residents. It did, however, have frequent visitors in the form of fishing trawlers and their crews. The boats, principally from the Philippines, China, and Vietnam, would periodically enter the lagoon to trawl some of the richest fishing grounds in the region and take shelter from heavy seas.

Beginning in April 2012, Chinese and Filipino vessels began a territorial game of chicken inside and around the Scarborough Shoal. At the time, I was chief of staff to the US ambassador to China Gary Locke, stationed at the US embassy in Beijing. Chinese activity in the shoal became a serious concern for US diplomats. Was China attempting a land grab in disputed waters at the expense of a US treaty ally? Virtually every day, new surveillance photos—taken by the P-8's predecessors, the EP-3 Orion reconnaissance aircraft and Global Hawk unmanned surveillance aircraft—tracked the Chinese ships in action. On April 20, three Chinese ships were at Scarborough with a fourth on the way. By May 11, ten Chinese ships, a combination of fishing trawlers and coast guard and maritime patrol vessels, occupied the shoal. Filipino vessels were outnumbered by more than two to one but holding ground.

Inside the embassy, US diplomats were monitoring Chinese activities closely. China's intentions seemed clear, and US opposition to those intentions was equally clear. While the United States does not take a position on the competing claims in the South China Sea, it staunchly opposes any unilateral attempts to claim territory there. Scarborough Shoal was particularly sensitive for the United States since one of the claimants is a US treaty ally. Moreover, the shoal lay within the Philippines' exclusive economic zone, which extends two hundred nautical miles off a nation's coastline. China's apparent attempt to take possession of the shoal despite US opposition was unprecedented, with troubling implications for the many other disputed islands in the region.

As it surrounded and harassed Filipino vessels in the Scarborough Shoal, China was applying pressure on Manila in other ways. Beginning in early May, China began to block fruit imports from the Philippines, one of the country's chief exports to China. Tons of bananas, bound for export, were left to rot in Filipino ports. China also began blocking flights to and from the Philippines, disrupting another significant source of income for Manila in the form of visiting Chinese tourists and businesspeople.

US diplomats and the Obama administration officials debated how forcefully to respond. Inside the embassy, one senior diplomat outlined a strategy: let China *think* the United States was backing down on Scarborough so Beijing would push its territorial claims there too aggressively, thereby further alienating Southeast Asian countries and chasing them further into America's corner. It seemed a remarkably risky strategy, giving ground at the expense of a US ally and setting a dangerous precedent for other contested islands.

By the middle of May, America's Filipino partners were growing more nervous. A US submarine was on a port call to Subic Bay—and the Philippines' foreign secretary wanted to make a visit to highlight the US-Philippines Mutual Defense Treaty. The US embassy believed Manila was also encouraging the Filipino media to publish stories on US submarine capabilities in the region. Manila was eager to remind Beijing it had a powerful ally.

Later in the month, Filipino diplomats sent further warnings to their US counterparts. China, they said, was trying to consolidate its position in the shoal. Chinese maritime surveillance vessels had by then essentially put the area under Chinese control. China was also increasing its economic pressure on the Philippines, calling in two large outstanding loans to the Filipino government.

Inside the embassy, the debate among US diplomats continued, including on the question of how far they believed China was willing to go. Some doubted that Beijing would further antagonize a close US ally, arguing that Beijing's steps so far were reversible. They noted China had not built any permanent structures on the shoal. Others warned that China's position was hardening.

On June 1, those warnings were borne out. Chinese vessels began to construct a barrier at the entrance to the shoal's lagoon. The barrier, composed of buoys and nets strung between Chinese vessels anchored there, blocked any more Filipino vessels from entering to resupply the remaining Filipino vessels anchored inside the lagoon. It was a bold move. The Philippines' then-president, Benigno Aquino, flew to Washington to register his concerns directly to President Obama. Six days later, satellite photos revealed that the Chinese vessels had laid a second barrier. Without the ability to resupply with fuel and provisions, the last Filipino vessel was forced to leave the shoal. In contrast, China, as one US diplomat noted to me, now had an "armada" of fishing trawlers inside the lagoon. Scarborough Shoal was effectively under its control.

Behind the scenes, Beijing and Washington were quietly negotiating a way forward. Inside the embassy, I remember concern but not urgency. US officials believed that China could be coaxed into reversing course—allowing the Philippines' fishing boats back into the lagoon and, more important, ending its attempt to take more formal possession of the shoal. This was a consistent pattern inside the State Department and the Obama administration at the time. Negotiation will work. China can be convinced. Let's not overreact. This approach applied to a whole host of issues and disputes between Washington and Beijing, including China's cyber activities against the US government and private sector. And yet even as this approach often failed to change Chinese behavior, it persisted.

For its part, Beijing insisted it was working toward the "joint development" of Scarborough Shoal with the Philippines and other regional claimants. Chinese diplomats said they were being "very restrained" in their response so far, while quietly warning they would not "fire the first shot" if the situation were to take a turn for the worse.

Later in June, the Philippines withdrew its last naval vessel from the seas around the Scarborough Shoal. Its withdrawal was the result of an agreement quietly brokered between Beijing and Washington. The Obama administration believed it had a commitment from the Chinese side to withdraw as well. However, while Beijing withdrew its naval vessels, some thirty Chinese fishing trawlers remained inside the lagoon. China's "little green men" were still occupying territory. A week later, twenty-six Chinese trawlers remained inside the lagoon. US diplomats in Beijing worried that China was reneging on its agreement and making its presence in the shoal permanent.

Word of the US agreement with China was spreading. Vietnam's ambassador to Beijing asked the US embassy if Washington had pressured the Philippines to back down on Scarborough.

Meanwhile, China was taking a harder line elsewhere in the South China Sea, protesting Vietnamese patrols near the Spratly Islands and urging Vietnam's National Assembly to avoid any action on a new maritime law that would formalize Vietnamese claims in the area.

China was drawing its own conclusions from the US response as well. Chinese diplomats noted US "moderation" in the Scarborough Shoal, which reassured Chinese leaders that the United States would not risk military confrontation over what the diplomats referred to as "trivial" matters.

In January 2013, the Philippines, frustrated by Chinese intransigence and US retreat, took Beijing to the Permanent Court of Arbitration, in The Hague, Netherlands. The case was brought to the court pursuant to the Annex VII of the United Nations Convention on the Law of the Sea, or UNCLOS. The move by Manila angered Beijing, which carried out a series of coercive actions intended to punish the Philippines. Among the moves, China added new restrictions on imports of Filipino fruit, again leaving tons of bananas to rot in port.

In July 2016, the tribunal would rule in favor of the Philippines, rejecting China's historical claims not just to the Scarborough Shoal but to all land features in the South China Sea. The unanimous decision was a stunning rebuke to China and its president, Xi Jinping, so comprehensive as to spark worries among its neighbors as to whether China would react with new land grabs in the region. With unusually strong language, the tribunal specifically cited UNCLOS, which had been ratified by both China and the Philippines, stating that the treaty "extinguished" China's claimed historic rights to the islands in the South China Sea. In its finding, the tribunal determined it "considers it beyond dispute that both Parties are obliged to comply with the Convention." Beyond the case of the Scarborough Shoal, the tribunal ruled that Mischief Reef, which China had by then almost fully militarized, was also in the Philippines' waters.[9]

China quickly found itself alone against its neighbors and the world. The United States, though it has not ratified UNCLOS, called on China to abide by the ruling.

"The world is watching to see if China is really the global power it professes itself to be," State Department spokesman John Kirby said, "and the responsible power that it professes itself to be."[10]

Vietnam, which has competing sovereignty claims in the South China Sea, also quickly endorsed the decision, a remarkable step for a country with a long-standing cooperative relationship with Beijing built in part on the countries' shared Communist Party leadership.

Despite the ruling, Scarborough Shoal remains under Chinese control. China has also not walked back its efforts to assert its "indisputable sovereignty" over the territory within the nine-dash line and has continued to deploy military installations on several land features claimed by multiple countries.

Today Beijing celebrates its success in the Scarborough Shoal to the point where Chinese academics such as Zhang Jie have employed the concept of the "Scarborough Shoal Model," a term first used in a May 2012 *People's Daily* article. The use of the term suggests that the Scarborough Shoal presents a model for study and application to other territories, despite opposition from the United States or the fact that such territories may in fact fall into other states' exclusive economic zones.

Erickson of the Naval War College singles out China's success— and the US failure to reverse it—as a defining victory in China's Shadow War on the United States.

"Beijing's reneging on a US-China-negotiated return to the

status quo ex ante in the 2012 Scarborough Shoal Standoff—and instead seizing the feature located within the Philippines' exclusive economic zone—is a success" for Beijing, said Erickson.

———————

C hina sent numerous warning signs about its ambitions in the South China Sea in the years prior to its blatant land grabs in the Scarborough Shoal and the Spratly Islands. Some of these signs were directed at its neighbors and others at the United States. In 1974, China dispatched ships to seize the Paracel Islands from Vietnam. It was a deadly encounter, leaving more than seventy killed on the two sides. The "Battle of the Paracel Islands" as it is now known left China with control of the islands, control it maintains to this day.

In the decades that followed, disputes in the South China Sea were largely dormant, with the exception of the 1988 Johnson Reef Skirmish with Vietnam and China's 1995 occupation of Mischief Reef, which precipitated skirmishes with the Philippines. Under Deng Xiaoping, China's policy was to "set aside disputes and pursue joint development." Still, South China Sea watchers did notice what Ian Storey referred to in 1999 as China's "creeping assertiveness" in the South China Sea—"A gradual policy of establishing a greater physical presence in the South China Sea, without recourse to military confrontation."[11] The United States, however, did not take notice until China's activities threatened the passage of US naval vessels through international waters.

In a remarkable confrontation in 2009, Chinese trawlers harassed the USNS *Impeccable*, a US surveillance ship. According to a Pentagon statement, the vessels steered directly into the path of the *Impeccable*, dropping pieces of wood in the water and using poles to try to grab the ship's acoustic equipment. Though the ships were not officially Chinese government vessels, the United States believes they were acting under the direction of Beijing, which frequently uses "fishing vessels" as a shadow navy. It was a dangerous, aggressive, and sometimes even comical standoff on the high seas.

"Because the vessels' intentions were not known, *Impeccable* sprayed its fire hoses at one of the vessels in order to protect itself," read a Pentagon statement. "The Chinese crewmembers disrobed to their underwear and continued closing to within 25 feet."

CNN quoted a Pentagon spokesman calling the incident "one of the most aggressive actions we've seen in some time. We will certainly let Chinese officials know of our displeasure at this reckless and dangerous maneuver."[12]

In 2014, China would take aim at another neighbor and competing claimant in the South China Sea: Vietnam. The "Hai Yang Shi You 981 standoff" bears the name of the Chinese oil platform at the center of the dispute. In May 2014, one of China's state-owned oil companies maneuvered the platform to within several miles of one of the disputed Paracel Islands. Vietnam immediately protested the rig's placement as a violation of its sovereign territory— and dispatched more than three dozen coast guard ships, tugboats, and fishing vessels to disrupt the rig's operations and prevent it from successfully moving into position. China responded with its own fleet of coast guard, surveillance vessels, and fishing trawlers. For weeks the two makeshift navies faced off in sometimes dangerous clashes. At least one Vietnamese fishing vessel was sunk. China withdrew the rig in August.[13]

Each interaction made clear that China was willing to use both military and nonmilitary assets to pursue its territorial goals, regardless of whether the United States or a smaller regional rival stood in the way. Oddly, in the case of the Hai Yang Shi You standoff, it was the smaller regional rival that appeared to have successfully thwarted China's land grab, whereas the United States—despite far greater military capabilities—was outplayed in both the Scarborough Shoal and Spratlys. In the Shadow War, asymmetric power can win.

———————

C hina's broad strategic goals in the South China Sea and farther afield are not—and have never been—secret. Andrew Erickson notes that the Chinese Communist Party (CCP) has enumerated—and methodically worked through—a series of national security goals since its founding in 1921, nearly one hundred years ago. Today it is on the verge of having achieved all of them.

China's security priorities begin with the party itself, with the core belief that the survival of the Chinese Communist Party is essential to national rejuvenation. That was achieved with the CCP's victory over the Chinese nationalist party, the Kuomintang, in 1949. From there China's national security goals are defined primarily by geography. Chinese leaders started by establishing unrivaled supremacy and control in the core Han-Chinese dominated heartland. From there they extended outward, securing stability and its own legitimacy in the minority-dominated borderlands of Tibet and Xinjiang, which China achieved in the 1950s. Once secure on the mainland, China looked to border disputes with its neighbors. This included fighting a bloody border war with Vietnam in 1979.

Once the Cold War ended, with China feeling secure in its own borders, Beijing turned its attention beyond the mainland, to what China refers to as the "Near Seas." These are the waters of the Yellow Sea, between China's east coast and Japan; the East China Sea, to the south of Japan and north of Taiwan; and the South China Sea, to China's south, bordering its Southeast Asian neighbors of Vietnam, Indonesia, the Philippines, and Malaysia. The Near Seas fall within what China refers to as the "First Island Chain," composed of Japan, Taiwan, the northern Philippines, and the Indonesian island of Borneo.

To achieve its ambitions in the Near Seas, China initiated a program of naval modernization under Chief Naval Commander Liu Huaqing in the 1980s. Though Deng had pursued a policy of setting aside these territorial disputes as China focused on economic growth, China began to pay increasingly close attention to the concept of maritime power toward the end of the Hu Jintao administration—shortly before the ramping up of activity in the South China Sea. In his final speech at the Eighteenth Chinese Communist Party Congress in 2012, the outgoing Hu called on China to become a "great maritime power" (*haiyang qiangguo*). The concept would be referred to more frequently in official documents and speeches under the subsequent reign of Xi Jinping.

China's policy in the South China Sea under the leadership of Xi can be characterized by increasing assertiveness and a heightened association of claimed territory in the South China Sea with China's core interests of sovereignty, territorial integrity, and national unification. The concept of a "great maritime power" is now more closely associated with China's "core interests" (*hexin liyi*)—essentially, nonnegotiable interests cited to justify specific policy positions. Since coming to power, Xi has referenced maritime power as part of his often-touted "China Dream" and even more recently called on the nation to realize its "great maritime power dream."

The increased focus on reclaiming what it sees as rightful Chinese territory under Xi is reflected not only in Chinese activity in the South China Sea but also in policies and legislation such as the 2015 National Security Law. Moreover, China declared an

Air Defense Identification Zone in the East China Sea in 2013, reflecting its renewed focus on the maritime sphere and territorial integrity more generally. The concept of a "great maritime power" reflects the grand strategic vision of the Chinese leadership and has manifested itself in concrete actions and policies in the South China Sea and beyond.

Beijing's ambitions include not only disputed islands in the South China Sea but also islands in the East China Sea, the Senkaku Islands (or *Diaoyu* in Chinese), which are claimed by Japan, and of course Taiwan, whose independence China still sees as illegitimate. They are both US allies that the United States is obligated to defend.

As its military capabilities grow, China is demonstrating these claims through airpower as well. As Admiral Davidson, commander of US Pacific Command, told the Senate Armed Services Committee in 2018, "In line with the Chinese military's broader reforms, Chinese air forces are emphasizing joint operations and expanding their operations, such as through more frequent long-range bomber flights into the Western Pacific and South China Sea."

Davidson made clear that these operations posed an immediate threat to US military influence and to deployed US forces in the Asia region.

"As a result of these technological and operational advances," he said, "the Chinese air forces will pose an increasing risk not only to our air forces but also to our naval forces, air bases and ground forces."[14]

China is now setting its sights on becoming a global military power. Despite China's near-term strategic focus on defending the Near Seas, in recent years Chinese leaders have developed and deployed naval forces capable of operating in the "Far Seas" of the central Pacific, the Indian Ocean, and beyond. These operations and capabilities are not about defending China but about projecting Chinese power abroad—and defending China's economic, diplomatic, and geopolitical interests around the world. In the simplest terms, China is building and deploying a true "blue-water" navy. Those expanded operations and capabilities have not gone unnoticed by the US military.

"While primarily a regionally focused military, China aspires to project power worldwide," Admiral Davidson testified before the Senate. "China's expanding global interests . . . have Beijing increasingly looking beyond the region."

To make that ambition a reality, China is building a global infrastructure to support a blue-water navy, much as the United States did in the late nineteenth and early twentieth centuries. Erickson says that China is expanding its access to foreign ports to pre-position support necessary for regular and sustained deployments in and around the Indian Ocean. This includes basing access in Pakistan and East Africa. China's participation in anti-piracy operations off Somalia beginning in 2008 were an early example of its ability to deploy and sustain navy vessels far from the mainland.

The spark for China's naval expansion came from the actions of the country China sees as its main adversary: the United States. In the 1990s, the United States demonstrated its vast US military capabilities in a series of engagements that led Chinese leaders and military commanders to determine they were gravely outmatched.

"Three major military events—Operation Desert Storm in 1991, the 1995–96 Taiwan Strait crises, and a US aircraft's mistakenly bombing the Chinese embassy in Belgrade, Serbia, on May 7, 1999, as part of a larger NATO operation—increasingly kindled a belief among Chinese leaders that America's technological superiority could enable it to attack Chinese assets with impunity," said Erickson.

Those perceived threats abroad, Erickson says, coincided with political changes inside China. At the time, then–Chinese president Jiang Zemin was expanding his authority. Jiang was a leader who prioritized China's military modernization—and he had the credibility and connections within China's military to realize those ambitions.

"This catalyzed heightened emphasis on funding and supporting megaprojects to develop such asymmetric 'assassin's mace' weapons as antiship ballistic missiles (ASBMs), as well as an upsurge in naval shipbuilding. The results are before us today," says Erickson.

———————

Like Russia, China wants to fulfill its growing ambitions—to win on each of these battlefields of the Shadow War—without actually going to war with the United States, that is, to "win without fighting." To do so, Erickson explains, China has developed a broad military strategy that scholars describe as "high-low." The "high" end is the fighting end: While China wants to avoid a shooting war with the United States, it must prepare for one as a deterrent to the United States and other adversaries. The "low" end encompasses hybrid warfare techniques, just below the threshold for war.

China's activities in the South China Sea illustrate the low end in action. China deploys paramilitary vessels and even nonmilitary vessels in the Near Seas to defend and sometimes invade or annex territory it claims as sovereign. They are China's seafaring equivalent of the "little green men" Russia deployed to Crimea.

China's coast guard and the People's Armed Forces Maritime Militia are, in effect, informal navies to complement the People's Liberation Army Navy, or PLAN. China also often uses fishing trawlers for such operations. Erickson explains that these forces give China options for gray zone operations, that is, operations to defend its various land claims and other maritime rights in the Near Seas while giving Beijing the ability to deny any involvement by the Chinese military. As with Russia, the Shadow War employs the power of the lie.

In 2009, it was Chinese fishing trawlers that harassed the USNS *Impeccable.* In 2012, when China first took control of the Scarborough Shoal, fishing vessels served as its frontline forces, only later joined by the coast guard and maritime militia. Again, in 2014, Chinese coast guard vessels, not navy warships, led the vanguard against Vietnamese ships in the Paracel Islands.

"In maritime gray zone operations," says Erickson, "Beijing employs its enormous coast guard and maritime militia to further its disputed Yellow, East, and South China Sea sovereignty claims using coercion short of warfare.

"China attempts to do just enough to further its goals," Erickson continued, "without triggering countermoves from the US, its regional allies, or other neighboring nations."

China's strong preference is to stay comfortably within the low end of its high-low strategy—to fight a Shadow War rather than a shooting war. However, to effectively do so, Beijing military planners believe they have to develop credible capabilities for a high-end conflict, even with the superpower United States. They want to send a message to the United States that it cannot easily win—and could very well lose—a war with China.

"To demonstrate ability to prevail in worst-case scenarios, and thereby also project deterrence in peacetime designed to pressure

Washington and its allies to accommodate Beijing's policy preferences without fighting," Dr. Erickson told me, "China is developing and deploying a 'high-end' counterintervention force."

That force is built around the military strategy known as "A2/AD," short for "anti-access, area-denial." The strategy is designed to deny access to an adversary by creating a massive kill zone for ships and aircraft close to China's shores.

"This A2/AD force is grounded in the world's largest regionally ranged, conventionally armed ballistic missile force," says Erickson. "It boasts potentially such 'game-changing' systems as two types of antiship ballistic missiles, a variety of antisatellite weapons, and even hypersonic technologies under development."

Today, Erickson notes, China's A2/AD capabilities extend from underwater to the surface into the air and all the way into space. It is a staggering expression of military power: the largest medium-range conventional missile force in the world (outnumbering China's nuclear-armed missiles by seven to one); new and more advanced antiship cruise missiles; antisatellite weapons, ranging from antisatellite missiles to so-called kidnapper satellites capable of snatching enemy satellites out of orbit; and two new military reconnaissance satellites to aid in missile targeting back on earth.

China has deployed its formidable missile force on land, ships, submarines, and aircraft to overwhelm the ability of US ships to defend against them. These missile systems give China the ability to target not just US ships and aircraft, but also US military bases in the region, including Andersen Air Force Base in Guam, Kadena Air Base in Okinawa, Japan, and any of the fifteen US Army and Air Force bases in South Korea.

China's objective is to demonstrate high-end military capabilities to the degree that it can deter the United States from even considering a high-end military conflict. Beijing aims to show

Washington that the potential costs of such a conflict, both in human terms and in military hardware, would be too great to even consider.

US commanders, while acknowledging China's remarkable advances, reject the very phrasing "anti-access, area-denial," because—they say—the United States can defeat China's systems and thus China does not actually possess the ability to deny the area to US forces. They prefer the term "counterintervention." They grant, however, that China at a minimum has increased the risk for US forces, especially carrier and carrier strike groups, which in today's environment make for very large targets. US commanders also express confidence in the ability of US submarines to continue to operate in these areas and—if necessary—to destroy China's A2/AD defenses to make way for the rest of the US Navy in the event of war.

However, China is making advances in submarines and submarine warfare as well. China's conventionally powered submarines, which have shorter ranges than nuclear-powered submarines and so operate close to the mainland, are becoming quieter and harder to detect. A noisy sub, as submariners say, is a dead sub. Crucially, these submarines can fire antiship missiles, designed with the US Navy in mind. US submarines maintain a clear advantage over Chinese subs, but US Navy commanders acknowledge that advantage is shrinking.

"Ultimately, this is a perishable advantage for the United States," Admiral Davidson wrote in April 2018. "Absent sustained, consistent investment and constant innovation, the PLA [People's Liberation Army] will catch the United States in this critical regime."

China's selection of weapons systems matches exactly with its strategic priorities. Erickson notes that—for now—the reach and capabilities of its missiles, warships, and aircraft diminish the farther one ventures from Chinese shores.

"China has placed the greatest emphasis on platforms and weapons systems most relevant to threatening any US or allied efforts to intervene in a Near Seas crisis or conflict," said Erickson. "These include, in particular, armament with large numbers of long-range precision strike weapons such as ballistic missiles (particularly land-based) and cruise missiles (land-, submarine-, surface ship–, and aircraft-based) capable of striking US and allied warships and regional bases."

This does not mean that China is limiting its territorial ambitions in the long term. In fact, China is developing weapons systems today that will allow it to expand its ambitions farther from its shores tomorrow.

"China's navy is developing—although not prioritizing intensively, at least for now—platforms optimized for long-range power projection such as nuclear-powered submarines, aircraft carrier strike groups, and long-range bomber aircraft," Erickson explained.

As China does so, it will come into more direct conflict with US submarines, aircraft carriers, and long-range bombers and, therefore, with US military influence. The "Far Seas" will be the next battlefield in the Shadow War.

For now the Shadow War is being fought in the Near Seas, which is already a direct challenge to US military influence in the region—influence the United States has maintained for decades virtually without challenge. American goals in the region do not include territorial expansion. The United States is not looking to establish any new colonies or build its own "unsinkable aircraft carriers" in the form of man-made islands. And, as a matter of policy, the United States does not take a position on any of the disputed islands in the South China Sea. Rather, US policy through Democratic and Republican administrations has been to uphold international law and keep trade routes open. US officials emphasize that such principles, backed by the might of the US Navy, have helped fuel Asia's economic boom, to the benefit of all parties—including China. The challenge now is how to continue to help keep the peace in Asia, even as China grows and flexes its military, economic, and diplomatic muscles.

So-called FONOPs, including regular flights over disputed territory by the P-8A Poseidon, are intended to establish in a tangible way that the United States does not recognize Chinese territorial claims and, therefore, that China's neighbors need not, either.

"To ensure stability, U.S. operations in the South China Sea—to include freedom of navigation operations—must remain regular and routine," Admiral Davidson testified. "In my view, any decrease in air or maritime presence would likely reinvigorate PRC expansion."

However, the United States has been conducting such missions for years and yet China has continued to claim new sovereign territory.

Some US defense experts see America's best hope under the waves. Today China already has the world's second most powerful navy. And, at least in purely numerical terms, it is on a pace to surpass the US Navy by 2030, a little more than a decade away. However, Erickson and US Navy commanders emphasize that US submarines retain a technological advantage that China is struggling—and will continue to struggle—to match.

Due to the US advantage, and China's continuing difficulty in overcoming it, Erickson and others again point to submarines as America's best hope to retain its edge. An essential part of the solution in the view of Erickson and senior naval commanders is to not only build and deploy more submarines, but to do so faster. Currently the US Navy is capable of building two *Virginia*-class attack submarines (the successor to the *Los Angeles* class) per year.

They believe the United States can accelerate that pace to three per year. The 2018 military budget called for adding fifteen submarines to the US fleet by the middle of the twenty-first century, raising the total deployed to sixty-eight. However, at the faster rate of three subs per year, the US Navy could achieve those numbers earlier, perhaps within a decade and a half.

In the event of war, *Virginia*-class attack submarines are capable of denying China's ambition of successfully operating a blue-water navy, by doing what those *Virginia*-class submarines are designed to do: hunting and destroying enemy subs far from home.

What may be essential, however, is far more comprehensive. Many US national security strategists believe the United States needs a strategic rethink designed for—and borrowed from—the Shadow War.

"The US now needs to pursue a 'high-low' strategy of its own," Erickson told me. "Deterring Chinese aggression by demonstrating ability to prevail in a traditional armed conflict and resisting Chinese maritime expansion constantly in peacetime."

For now, America's high-low strategy relies on the same tactics and power projection as it has for decades: maintain an overwhelming military advantage in the event of war, while sailing and flying through contested waters to undermine China's territorial claims.

———

In August 2018, three years after our exclusive P-8 flight over China's man-made islands, the US Navy invited CNN back on board the P-8 for another mission over the South China Sea. The Chinese navy again challenged the US flight crew repeatedly, demanding the P-8 leave the airspace. And the US flight crew again refused, reading from a familiar script reasserting the US government position that this is international airspace over international waters. These communications were calmer than they had been on our trip in 2015. CNN senior international correspondent Ivan Watson, who was on board for this mission, told me they seemed to have become "ritualized."

"The pilots read scripts in response to what sounded like scripts read by the Chinese," Watson told me. "It had become a new normal. It's like they were doing a dance."

Still works in progress in 2015, the islands were now fully militarized military outposts. Visible from the air on Fiery Cross Reef were radar towers, power generation plants, and buildings as tall as five stories to house military personnel. The runways—long enough to accommodate virtually every aircraft in the Chinese military—were complete. On nearby Subi Reef, the Poseidon's crew counted some eighty-six vessels deployed in the island's deep, man-made lagoon, including Chinese coast guard and navy ships—an enormous contingent for just one of several of the man-made islands in the South China Sea. What were missing, however, were people. One of the flight crew told Watson the crew didn't see more than a dozen people on one of the islands. It wasn't clear if they were hiding from a US aircraft their radars would have been monitoring from miles away, or if they simply weren't there in numbers.

"It felt like a Potemkin village," Watson told me.

Increasingly, however, China's man-made islands were hosting full-fledged military operations. In May 2018, the United States detected the deployment of antiship and antiaircraft missiles on three of the man-made islands during Chinese military exercises. The missiles form an integral part of China's anti-access, area-denial strategy, which is itself designed with the United States in mind. Beijing's blunt message to Washington appeared to be: we are prepared to make these waters unsafe for US warships. President Xi's promise to President Obama in 2015 not to militarize the islands had become meaningless.

Chinese warnings would not end there. Later in the same month, the People's Liberation Army Air Force announced that several bombers had successfully landed and taken off from an island in the South China Sea, later identified as Woody Island in the Paracels. The bombers included the nuclear-capable H-6K, which is designed to attack US carrier groups and land targets at a range of more than two thousand miles.

US Pacific Command took notice, and a spokesman criticized the exercises as part of "China's continued militarization of disputed features in the South China Sea." China's Foreign Ministry, as per its normal practice, dismissed US objections as an overreaction, saying in its own statement, "The islands in the South China Sea are China's territory. The relevant military activities are normal trainings and other parties shouldn't over-interpret them."[15]

The "facts on the sea" stood firmly in China's favor. China had built the islands, militarized them, and made them operational— all over the vocal objections of Presidents Obama and Trump. Both administrations continued freedom-of-navigation operations to make US opposition clear. But those standoffs on the surface and in the air failed to change Chinese behavior or reverse Chinese territorial gains. China had won this battle of the Shadow War.

LESSONS

China's construction of man-made islands in the South China Sea is an expansive and imminently tangible challenge to the rules-based international order established and championed by the United States since the end of World War II. It is an enormous land grab in the midst of territory contested by more than half a dozen nations, several of which are US allies, and one that shatters the international treaty governing the sea. Moreover, unlike Russia in Ukraine, China accomplished this land grab, in the process redrawing the borders of Asia, without firing a shot. In doing so, China has successfully challenged America's decades-old role as the dominant military power and keeper of the peace in Asia. It is an alarming and sobering precedent for other ongoing disputes between Washington and Beijing in the region, including those over other contested landmasses in the South and East China Seas, such as the Senkakus, claimed by Japan, and the Scarborough Shoal, claimed by the Philippines. Notably, both Japan and the Philippines are countries to which the United States is bound by treaty to defend against foreign aggression.

As with Russia's aggression in and around Europe, China signaled its territorial ambitions in the South China Sea over decades, signals repeatedly missed or downplayed by successive US administrations of both parties. Throughout, American protests to Beijing fell on deaf ears. As a result, the United States enabled China's land grab by repeatedly failing to back up its protests with sufficient costs, or credible threats of force. Today China's man-made islands are "facts on the sea," so to speak, that even the most aggressive China hawks in the United States acknowledge will almost certainly remain no matter what signals the United States sends now.

CHAPTER 6

War in Space

(RUSSIA AND CHINA)

I n May 2014, a small team of airmen at the Joint Space Operations Center, or JSpOC, at Vandenberg Air Force Base in California noticed something they'd never seen before. Russia had launched a satellite the month before—one of dozens of commercial and military satellites launched every year. This particular Russian rocket lofted a communications satellite into space and what appeared to be the usual collection of space junk—ranging in size from massive spent rocket stages down to chips of paint. Much of that debris either falls back toward earth over the ensuing days and weeks, burning up safely in the atmosphere, or joins the growing clouds of space garbage in orbit, each piece tracked daily to avoid collisions with satellites or—more important—with manned spacecraft, including the International Space Station. A few weeks after that 2014 Russian launch, however, one piece of that "junk" came to life.

The airmen at JSpOC[1] watched over the next several days as the unidentified space object made eleven close passes of the rocket's upper stage. It was an elaborate space dance, possible only if the object was equipped with both thrusters and sufficient fuel to maneuver through orbits and to do so with enormous precision. This is a capability consistent with antisatellite weapons, or "kamikaze satellites," that is, satellites with the ability to maneuver up to other satellites in order to interfere with, disable, or destroy them. The satellite was cataloged as "Kosmos 2499"—a name corresponding to the 2,499th space launch by Russia—and designated for closer observation. As it would turn out, Kosmos 2499 had only begun its journey.

JSpOC—referred to as J-Spock in the space community, the *Star Trek* references not lost on anyone—looks more like a corporate call center than the bridge of the USS *Enterprise*. Concentric circles of identical desks with computer screens are manned by airmen and -women wielding computer mice and touch pads as they scan the skies for potential threats. Their screens can give them a vivid, three-dimensional rendering of objects in space down to the size of a softball. Just a click or two allows them to zoom in for inspection. This is the look and feel of space warfare: computer screens as battle stations, cavernous, temperature-controlled ops centers as the new front lines.

At one desk sat First Lieutenant Andrew Engle, one of a number of newly designated Defensive Duty Officers, or "DDOs," charged with monitoring space launches for potential danger. Engle watched his screens with the alertness of a sentry guarding a forward military outpost. Despite the immense distances between them and the satellites they are tracking, the airmen take threats in space as seriously as an infantryman scanning the next hilltop for a sniper, or a fighter pilot searching the horizon for enemy aircraft.

Engle's bogeys, however, are hurtling through space hundreds of miles above his head at speeds of 17,500 miles per hour. Kosmos 2499 got his special attention.

"It wouldn't be on the same plane and orbit, unless it was intentional," Engle said. "This is something that is on the new frontier of space that we're seeing from our adversaries. It's highly technical, highly skilled, and it's something that we're definitely, obviously watching closely to see what the capabilities are."

Engle is assigned to the 614th Air and Space Operations Center, a team charged with defending military satellites and which is one small part of the larger Air Force Space Command. Long before President Trump ordered the creation of a "space force" in 2018, Air Force Space Command was already a fully functioning wing of the US military, with more than 30,000 military and civilian employees worldwide, a $9 billion annual budget, six bases, and 134 locations around the globe, though it operates largely out of sight and out of mind of the American public.[2] Until recently, much of Space Command's operations were classified. That has changed as military leaders feel increasingly compelled to warn the public and political leaders of the urgency of the emerging threat to US space assets.

As DDO, Lieutenant Engle keeps especially close watch for any threats to the country's highest-value space assets, including surveillance satellites, GPS satellites, and—crucially—the nation's network of nuclear early-warning satellites. The technology at his fingertips looks easy and straightforward. Like a video game, satellites and other spacecraft show up in 3-D, computer-generated images. But while the images are artificial renderings, the positions and flight paths are real, beamed down from space in real time. Sitting at his station in the state-of-the-art ops center, he showed me a simulation of past maneuvers by a Russian kamikaze in flight. Like an assassin stalking its target, it repeatedly orbited a US satellite within a thousand yards—frighteningly close in space, considering both objects were traveling at twenty times the speed of sound.

"As the US satellite—we call it the blue satellite—is going around in space, the Russian craft—we call it the red satellite—is mirroring everything the blue satellite does," Engle said. "Due to the nature of space, that is not something that we see just happen by chance."

Kosmos 2499 performed several "orbits" of the US satellite before firing its micro-thrusters to move on to its next target. From such distances, it could disable or destroy a US satellite in a number of ways. Russia and China have experimented with lasers and other directed-energy weapons that can "dazzle" satellites with relatively low-powered electromagnetic interference. This is the space equivalent of distracting a commercial pilot with handheld laser pointers: disruptive and potentially dangerous, but short-term and reversible. More powerful directed-energy weapons can permanently disable satellites, can "fry" them with a more powerful burst of energy.

More daunting is the brute force of ramming satellites like a bullet, shattering them into thousands of pieces and littering low-earth orbit with debris. The movie *Gravity* portrayed an accidental collision between a space shuttle and a storm of space junk. A real collision might be impossible to capture on film since those pieces of junk, depending on the angle between their orbits, could be traveling at speeds too fast for the human eye to register. The destruction would then be instantaneous.

Navigating this orbiting minefield is the intense concern of military and civilian space operators alike. Each satellite is a precious collection of technology and capability, costing tens of millions of dollars to manufacture and tens of millions more to launch into space orbit. "Space is hard," space operators like to say. No one wants space to become inaccessible.

Space is a new and dangerous front in the Shadow War. Russia, China, and other US adversaries are rapidly developing and deploying offensive capabilities in space designed to undercut the United States' enormous advantage in the space realm, while exploiting the US military's and US civilian population's unmatched dependence on space assets and technologies. And like many fronts in the Shadow War, though the threat is growing, the United States is still debating how best to respond.

———

Kosmos 2499 is not a unique threat. Today at least four satellites—two launched by Russia and two by China—are doing things no other man-made space object has ever done before. Analytical Graphics Inc., or AGI, has been tracking these space objects since their launch. All told, AGI's Commercial Space Operations Center, or ComSpOC, in rural Pennsylvania is tracking some ten thousand objects in space, in effect crowdsourcing a global network of radar antennae and telescopes to create a virtual, 3-D picture of space traffic in every orbit down to objects just ten centimeters in size. ComSpOC (pronounced "Com-Spock," again with *Star Trek* in mind) serves as a kind of space air traffic control for the hundreds of governments and private companies operating satellites today. Operational satellites account for about 1,500 of the objects ComSpOC tracks. The rest are debris, ranging from decommissioned satellites to spent rocket stages to smaller satellite pieces and components that are the space equivalent of the tire treads, hubcaps, and broken glass littering the shoulders of earthbound highways. AGI's visual rendering of the objects orbiting the earth resembles a porch light swarming with thousands of mosquitos.

In 2019, the US military plans to begin operating the most capable space monitoring system ever deployed. Dubbed the "space fence," the state-of-the-art ground-based radar located in Kwajalein Atoll in the Marshall Islands will be able to track more than 100,000 objects down to the size of a softball. One commander described it to me as a giant searchlight scanning virtually every object orbiting earth. And today that searchlight is picking up more and more potential threats.

"The United States has taken a position for many years that there should be no fighting systems in space," Paul Graziani, the CEO of AGI, told me. "Fighting a war in space is something that every US government official I've ever met doesn't want anything to do with. Unfortunately, our adversaries have driven us in another direction. If these weapons are ever used, if we're not already in World War Three, we will be there in no time."

Why World War III? US military commanders believe space weapons are intended to give US adversaries the ability to do catastrophic damage to the US homeland and the US military. In their view, space weapons are weapons of all-out war.

At ComSpOC, satellite technicians watched in real time as a second Russian satellite, dubbed "Luch," took offensive threats in space to a new level. Launched in 2014, just a few months after Kosmos 2499, Luch has not one, but two dangerous capabilities. Like its predecessor, it can move through orbits, from one foreign satellite to another, coming dangerously close to observe, disable, or destroy.

Moving within an orbit is not new. Most satellites have some propulsion to make small adjustments in flight. But to move from one orbit to a completely different orbit, which can mean travel of hundreds or thousands of miles, or to sidle up close to another satellite and circle it, which means numerous adjustments at massive

orbital speeds, requires much more power, fuel, and remote pilot-
ing skill. And those are the advantages of maneuverable satellites
such as Luch. To accomplish that, it relies on both old-generation
gas thrusters and new-generation electromagnetic propulsion, in
effect shooting a beam of electricity to maneuver through space.
This combination gives the Luch the speed and maneuverability of
a space fighter jet. At the same time, it operates as an orbiting platform for spying, able to intercept the communications going to and
from some of the most sensitive satellites in space, hoovering up every byte of information with a massive space antenna resembling a
rigid, circular fishing net. The Luch has the ability to sidle up next
to some of the most sensitive US surveillance and communications
satellites to conduct the space equivalent of tapping the phone lines
at CIA headquarters in Langley, Virginia.

Over the course of a year, the Luch would cozy up to three US
satellites handling some of the most sensitive military and government communications. As we watched inside ComSpOC, it was
shadowing a commercial communications satellite.

"Having their satellite cozy up next to other satellites does af-
ford the Russians opportunities that they would not have had from
anywhere else," says Graziani. "Sitting close to those satellites allows Luch to intercept signals that were intended to go solely to the
targeted commercial communications spacecraft."

This capability is based on simple physics. The beams coming
from the ground station toward the target satellite are fairly narrow,
so in order to intercept those signals a satellite needs to be fairly
close to where the beam is aimed. Some of those streams going up
are encrypted, some are not. And even the encrypted streams aren't
impossible to decrypt given enough time and computing power.

In 2013, shortly before Russia's launch of Kosmos 2499, China
introduced a new kind of space weapon as well. Observers initially

assumed the Shiyan-7, which translates as "Practice, number 7," was a conventional satellite. Then, through a series of complicated movements through orbit, it put itself in position close to a companion satellite launched at the same time, the Chuangxin-3. It seemed to be experimenting in approaching, circling and shadowing the Chuangxin, demonstrating Shiyan has a capability similar to Russia's Kosmos 2499 and Luch satellites.

Later, in 2014, AGI technicians noticed that the second satellite had disappeared from their screens. Had it been destroyed? There was no sign of debris. Had it docked with its larger partner? It wasn't clear. Earthbound sensors can accurately determine the location and size of space objects, but they can't always discern their outlines.

Throughout 2014 ComSpOC watched as the smaller satellite reappeared, and then disappeared again. AGI analysts could see only one explanation: Shiyan-7 was grappling and then releasing its smaller partner, "practicing" an entirely new capability. Paul Graziani and his team dubbed the Shiyan-7 the world's first "kidnapper satellite."

"It is extremely maneuverable and has performed a number of missions," he told me as we watched the maneuvers unfold on ComSpOC's massive screen. "Going close to a small satellite it launched, grappling and then releasing and then grappling again, back and forth, it grabs and releases."

Chinese state media reported the satellite's robotic arm was designed to retrieve space debris and carry out satellite maintenance. However, US space commanders see more threatening applications. China makes no secret of its military plans for space. Deploying weapons in space fits into China's broader military strategy of "fighting and winning local wars under informationized conditions." In layperson's terms, that means applying information technology in all aspects of military operations, from cyber warfare on the ground to disrupting and destroying an enemy's information technology in space, including by targeting satellites. In such a strategy, China is David and the United States is Goliath: the most advanced, most dependent on space and information technology, and therefore most vulnerable to attacks on those technologies.

"We would absolutely be shocked if the US military were not on a war footing now based on what we see," said Graziani. "There is no doubt that both Russia and China have seen weaponizing space as a way they could asymmetrically shift the odds in their favor."

In an era when US commanders are increasingly focused on asymmetrical threats from a whole range of state and nonstate actors, space provides a new opportunity for adversaries big and small to diminish or eliminate US technological advantages.

China and Russia are taking the lead in weaponizing space. But they are not alone. Iran and North Korea are also experimenting with laser weapons directed at space. And, in theory, any country with a space program could weaponize—and very quickly. AGI estimates that some one thousand objects in space have at least some maneuverable capability. And in space, when objects travel at 17,500 miles per hour, every maneuverable object is potentially a space weapon.

"From a military perspective, if something is maneuvering it could potentially come and take you out," said Graziani. "If you wanted to use your car as a weapon, you could. You could injure or even kill people with a car. Same thing with any satellite. Satellites move at such tremendous rates of speed that a satellite running into another satellite would destroy it."

In space, where a single object just two centimeters across carries the force of a speeding SUV, neither size nor numbers are barriers to power. During World War II, bomber raids might involve a thousand bombers and thousands of airmen. In space, weapons in the single digits can devastate.

For decades space had been a benign environment. Even during the height of the Cold War, the United States and Russia agreed to keep weapons out of earth orbits. Competition in space was certainly competitive, as the race to the moon demonstrated, but not overtly hostile. Then, when the Soviet Union collapsed, space was no longer even competitive for the United States. It had lost its only credible competitor. In the last decade, however, with the rise of China and reemergence of Russia as a power with global ambitions, space is once again both competitive and dangerous.

———————

The US military has staged numerous simulations of war in space and each paints an alarming picture. In the worst cases, attacks in space would presage an all-out, even nuclear war on the ground. Even a more limited engagement in space could have devastating consequences for the US civilian population and the US military.

For Americans, the first space war would begin without a sound. A coordinated series of cyberattacks races through the United States at the speed of light. Televisions go dark. Internet connections are unusually, dramatically, painstakingly slower. ATMs start to malfunction. Early on, it might seem like a series of unfortunate cyber glitches. There is little immediate alarm.

The front lines of the battle then extend from the cyber realm into the far reaches of space. Ground-based lasers target US communications satellites in low-earth orbits. Missiles launched from enemy ships and aircraft destroy the GPS satellites circling more than twelve thousand miles above earth. Past that, thousands of miles higher, in geostationary orbits—the "holiest" orbit of all, as some space commanders say—newly deployed kamikaze satellites disable the country's most essential nuclear early-warning and surveillance satellites. In a worst-case scenario, a broad-based attack creates enough wreckage to render orbits unusable for years.

With the loss of satellites, the effects in the civilian world become more comprehensive. The financial markets—with trades dependent on time hacks provided by the military's constellation of GPS satellites—have been paralyzed and shut down. The internet stops altogether. Business comes to a halt, as credit cards and bank machines become useless. Mobile phone services, already patchy, fail completely. Like the moment the second plane hit the twin towers on 9/11, there is a realization the country is under attack.

The loss of GPS satellites causes disruption far beyond the financial markets. Traffic lights and railroad signals—also timed by GPS—default to red, bringing transport to a standstill. Air traffic is suspended as pilots lose navigation. The loss or disruption of NASA and NOAA satellites means no more weather forecasts.

Disruption of the nation's power grid and water treatment plants soon follows. Basic services are suffering. With fears for public order, government officials consider a state of emergency.

Peter Singer painted a similar picture of the first hours of such a space-age conflict in his 2015 novel *Ghost Fleet: A Novel of the Next World War.*[3] Such a war would see America come to a screeching halt. Our increasingly connected world would suddenly become disconnected. America's greatest strength would have become its greatest vulnerability.

A space war would be confusing for civilians. For the military, it would be paralyzing, blinding service members on land, air, sea, and below the surface of the waves, in the process disabling a whole host of modern and sophisticated American weaponry.

"We'd go back to the way we fought in World War Two," says

General William Shelton, the former commander of US Air Force Space Command. "Think of all the things that won't exist without space—remotely piloted aircraft, all-weather precision-guided munitions. Now we can target anyplace on the planet, anytime, anywhere, any weather. That will be lost."

With the GPS system down, the United States would no longer be able to position its drones over ISIS targets in Syria or Al Qaeda in Pakistan. US cruise missiles and smart bombs would go off target. US ships and warplanes would need to revert to paper maps and radio communications. US soldiers in combat would lose visibility of enemy fighters.

"The foundation of almost any military operation these days is based on some sort of space capability," said General Shelton. "Whether it's communications, or GPS, or intelligence capability that's provided from space. All that information, and let's face it, this is an information era that we're in today."

The first hours of the space war could render useless the weapons of the largest military and intelligence apparatus ever assembled, while leaving the US military unable to strike back.

"It's not a force-on-force kind of thing. It's the person with the best information is likely to win," Shelton explained. "So, all that information comes down via space and that's the way we've—that's become the American way of war, very information heavy, very intelligence heavy. If you take all that away and we go back to industrial-age warfare, I don't know that we know how to operate that way anymore. So, to say that a space war might be devastating to the United States military, I don't think that's any exaggeration at all."

With a successful campaign in space, a much smaller and less powerful rival could quickly level the playing field with the United States. The pace and power of the Shadow War can be frightening.

A series of war games like this conducted through 2015 "did not go well," as one senior intelligence official told me. The exercises would serve as a wake-up call to the US military, sparking an unexpected call to arms at an unusual meeting of the military's top space commanders in the spring of 2015.

The Space Symposium, held in April at the upscale Broadmoor hotel in Colorado Springs, Colorado, is normally a dry, forgettable event. The annual gathering of government and industry groups resembles other conferences in the defense arena: a chance to network, and to buy and sell the most advanced weapons technology.

But the April 2015 symposium would turn out to be something very different. The guest of honor was Deputy Secretary of Defense Robert Work, who, as Secretary of Defense Ash Carter's understudy, led America's space efforts. That year in Colorado, Deputy Secretary Work called together America's most senior space commanders, as well as other industry experts with security clearances, in classified session. Though his full remarks were secret, Work's staff released a declassified summary that hinted at his message.

"While we rely heavily on space capabilities, in both peace and war, we must continue to emphasize space control as challenges arise," he told the audience. "To maintain our military dominance, we must consider all space assets, both classified and unclassified, as part of a single constellation. And if an adversary tries to deny us the capability, we must be able to respond in an integrated, coordinated fashion."[4]

Some of those present described his warning in far starker terms. Paul Graziani, the CEO of Analytical Graphics Inc. (AGI) was among them. He recalled, "It was the most interesting meeting I've ever been in in my entire career. There were probably one hundred and fifty people or so in the audience at the time: military people, intelligence community folks, and contractors. And Secretary Work really laid out what the threat really was and what we're going to do about it."

In the audience, a primary target of Work's warning was General John Hyten, then head of US Space Command. Hyten himself told me he took it as nothing short of a dressing-down for him and the rest of US space forces.

"We did not move. It was really April fifteenth of 2015 when we started to move. And that was a day when the deputy secretary of defense got an airplane, flew out here, and kind of looked and said, 'Are you ready if a war extends into space?' And the answer was, well, not really. And he said, 'Well, what does it take to get ready?' And I said, 'We know how to do it. We just need a little bit of resource and a little bit of time and we'll do that.'"

Work warned that the United States was simply not prepared for war in space, but he ordered commanders to get prepared immediately. The urgency of his words resonated throughout all facets of Space Command. The US military is only now catching up to this new, daunting reality. The United States clung too long to its assumption that space was safe—that the rules established in the 1960s in the midst of the first space race still hold today. That outdated assumption led to a series of other mistakes, commanders say, including neglecting to defend space assets from threats and refraining from testing and deploying US weapons in space, at a minimum, as a deterrent.

Like a space-age Paul Revere, General Shelton had been sounding the alarm for nearly a decade before that April 2015 meeting—warning his commanding officers that the United States must radically adjust its approach to space, beginning with abandoning the idea that space is uncontested territory.

"You're talking to a guy that was looking to really precipitate action and action didn't start quickly enough," he said. "Could we provide active defense of our satellites? The answer is no."

Even China's 2007 antisatellite test had not sparked immediate, definitive action. What held the United States back? General Shelton says the natural inertia of a giant military bureaucracy is partly to blame.

"That's a long wait," General Shelton told me. "You know, the US government is very capable, but very large, and it just takes a long time to move that aircraft carrier, so to speak, in the right direction."

But there was something more. US space forces had come to believe in their own invincibility. Military history is riven with surprise attacks and setbacks that exposed deep and dangerous weaknesses.

Pearl Harbor demonstrated the US Navy's vulnerability to coordinated air attack. The Iraq War demonstrated US ground forces' vulnerability to insurgencies. The rise of ISIS demonstrated the vulnerability of the Obama administration's reliance on US-trained indigenous forces from Iraq to Afghanistan, Africa, and beyond. Russia's invasion and annexation of Crimea demonstrated NATO's vulnerability to asymmetric, hybrid warfare. US space forces had, fortunately, not yet suffered their own space-age Pearl Harbor. But those Chinese and Russian advances in space weaponry were signaling that such an attack was possible or even probable, without significant change in US strategy and resources.

"The overall message that Secretary Work was delivering was that our adversaries have decided that space is one of the ways they are going to come at us," added Shelton. "We had hoped that space would remain a sanctuary and people would not militarize the operations in space but that hasn't turned out to be the case. And so now we are going to make a large effort to defend our own assets and cause the enemy to lose their space capabilities." Deputy Defense Secretary Work had put his space forces and the US military as a whole on a war footing in space for the first time.

The April 15, 2015, meeting generated ripple effects throughout the US military. Commanders across the armed forces were learning that the technology they relied upon on the battlefield was more vulnerable than they realized. How would they fight without space? And, crucially, would they lose without it? Inside US space forces, commanders were put on notice that they had to adjust to the new threat environment and do so quickly. Change meant doing a whole range of things better: identifying space threats, defending space assets, and, if called upon, developing offensive capabilities to shoot back at adversaries in space. With that reality in mind, the United States has been standing up its space forces at installations across the globe.

Visit Schriever Air Force Base in Colorado Springs, Colorado, and you'll notice something is missing. Schriever sits in the broad plains stretching east from the Rockies, with Pikes Peak as an imposing backdrop. On-site, it looks like dozens of air force bases scattered across the United States, with bunches of satellite antennae ensconced in protective domes (the "golf balls," as they call them) around utilitarian, low-rise buildings housing the command center, operations center, and housing as well as the requisite fitness center and preschool for service members' children. The winds are so strong here that airmen run their annual fitness tests indoors to avoid help or hindrance from the wind. High fences topped by barbed wire circle the base grounds, with another, more imposing perimeter around the restricted buildings that are home to its most sensitive mission teams.

What Schriever is missing is something every other air force base—naturally—is built around: a flight line, that is, a runway and aircraft. The 50th Space Wing that calls Schriever home doesn't have any fighters, bombers, or surveillance aircraft. They don't own or operate a single plane. All the "birds" they fly are hundreds and thousands of miles above them. And those "birds" and their pilots are now on a war footing.

"We have adopted a war fighter culture in space now," said Lieutenant General David Buck, commander of Air Force Space Command.

General Buck has the bearing of a field commander: no-nonsense, direct, and ready to brawl. For him and his team, space war is not a distant, theoretical threat. It is as real as ISIS in Iraq or Russia in Eastern Europe.

"I think if you came to space ten years ago, space was seen more as a service, a service provider, if you will," Buck said. "Well, no more. We're war fighters at Air Force Space Command, and that's pretty cool."

"Master of Space" is the 50th Space Wing's nickname, emblazoned on unit patches under a rendering of the opinicus, a flying griffin-like creature dating back to medieval times—a thousand-year-old mascot to represent a twenty-first-century space-fighting force. Today's nickname is actually a variation on the unit's original dating from World War II: "Master of the Sky." Back in the twentieth century, its pilots flew air cover for the Normandy invasion; later, nuclear-armed fighter-bombers based in Germany during the Cold War; and finally, the first air force missions dropping precision-guided bombs during the Persian Gulf War. But in 1992, they flew their last aircraft and were re-tasked to "fly" seventy-eight of the nation's most sensitive military satellites.[5]

Today, if it were a country, the 50th Space Wing would rank

sixth in the world for satellites under its control, just behind India and ahead of the European Space Agency. You'd think that mission would require an enormous number of personnel, or "space pilots," as it were. The truth surprised and somewhat alarmed me.

The third floor of the wing's command center is home to half a dozen command modules, known as "MODs," housed behind the kind of thick metal doors you see on bank vaults. In each MOD, a squadron of airmen and -women fly, control, and protect an entire constellation of satellites. Behind one door, a team flies the military's four "MilStar" satellites, which provide secure communications to US war fighters globally. Behind another, another team flies its "EHF" constellation, providing extremely high-frequency communications, reserved for the most secure and secret military and intelligence communications. These are the satellites the president uses to communicate the most sensitive commands to deployed forces. EHF comms are also particularly popular with US special forces. At the end of the hallway, another squadron flies the military's so-called neighborhood watch satellites. These are the satellites tasked with monitoring space for any potential threat to US space assets, such as Russia's Kosmos 2499.

The team flying more satellites than any other is 2nd Space Operations Squadron, or "2 SOPS," which is responsible for the Global Positioning System, or GPS, constellation. Now approaching the thirty-eighth anniversary of the launch of its first satellite, the GPS constellation is the biggest and arguably the most important network of satellites in space today. Most people know of GPS for its positioning capability that is the basis of every satellite navigation system. And that mapping function is crucial for civilians and the military alike. Every aircraft, every naval vessel, every submarine, every war fighter, every guided munition, every drone depends on GPS mapping. But GPS also provides precision timing, transmitting time stamps accurate to the millisecond globally. Today everything from bank transactions to stock trades to traffic lights depends on GPS time stamps to function. And this dependence extends far beyond the United States. Every country in the world uses GPS. Space Command estimates some four billion people around the globe take an action dependent on GPS every day. It is a massive, state-of-the-art technology provided to the world by the US military for free.

Captain Russell Moseley is crew commander of the 2nd Space Operations Squadron stationed at Schriever. He and his team's command module felt oddly antiseptic and silent, separated from threats and potential adversaries by hundreds of miles of sky. But in the view of the 50th Space Wing, that separation is artificial. Objects move at lightning speed in space. Hundreds of miles disappear in seconds. The dangers may look distant, but they are immediate and real.

"Across the Fiftieth Space Wing, everybody is on alert," Moseley said. "It's part of wearing the uniform."

Today, there are twenty-four active GPS satellites, with a further ten reserve satellites guaranteeing global coverage twenty-four hours a day, seven days a week. And yet, inside 2 SOPS's MOD, flying those thirty-four satellites falls to a tiny team.

"How many folks on duty here?" I asked Captain Moseley.

"We have seven military on duty right now. And one civilian contractor on duty," he said.

"And thirty-four GPS satellites you're in charge of," I said.

"Yes, thirty-four GPS satellites flying around the globe, providing GPS twenty-four/seven to the world."

That's a ratio of more than four satellites, to each service member on duty—four $250 million satellites, essential to running communications, navigation, and, increasingly, any networked device.

Stop the satellites and you stop our twenty-first-century wired world. Not long ago, those satellites and the space they occupy were considered safe from everything but asteroids and space junk. No longer.

"We currently fight in a contested and degraded environment. And that is only going to become more so in the future," Captain Moseley told me. "Space is the forefront for everyone across the globe. It's the next frontier."

It is also a new front in the Shadow War.

Not too far away from Schriever, the 460th Space Wing at Buckley Air Force Base in Aurora, Colorado, arguably has the most urgent task: protecting America's nuclear early-warning satellites. Colonel David Miller commands the 460th.[6] And like many of his colleagues, he was educated on the battlefield. Miller served in Baghdad as military advisor to the Iraqi prime minister and minister of the interior, developing strategies for Iraqi security forces to fight ISIS and other terror threats. For him and others, space war is not a theoretical threat but a threat just as real as ISIS suicide bombers and snipers threatening US and Iraqi forces on the ground today.

"Trajectories will vary, but rough order of magnitude, you're looking at about thirty- to forty-minute flight time of an intercontinental ballistic missile," said Colonel Miller. "That may sound like a lot of time. But when you are providing warning and trying to inform decision makers on what the response should be, you could imagine that the expectation for us, as the first detector and first reporter of that information, is significantly short."

Today, however, the early-warning system mostly relies on just four satellites circling the earth in geosynchronous orbit, some 22,000 miles up. Losing just one of those satellites would significantly damage US visibility, which is why the United States took notice when China completed another suspicious launch: sending a maneuverable satellite, like Russia's Kosmos 2499, some 18,000 miles up, within striking distance of geosynchronous orbit and those early-warning satellites. US military analysts see this launch as a test for conducting space warfare in the highest earth orbits. This is, in its own right, an escalation. China and Russia are demonstrating that they are developing weapons that can threaten US space assets in every orbit, even the most distant and most crucial orbits where America's most sensitive satellites fly. That is a reality that US military commanders find—in a word—unacceptable. If adversaries can target and take out the satellites protecting the United States from nuclear attack, those space weapons themselves are by definition an existential threat.

Vandenberg Air Force Base in California, nestled north of Los Angeles in the middle of wine country made famous by the movie *Sideways*, is another space unit preparing for war. General Buck's command post is hidden from sight in an old hangar. The hangar has a deep history in the US space program: it served as a mating facility for the giant Atlas rockets that powered some of the first space launches. Now Buck and his team of "space warriors" are developing the strategies and weapons for a new space race.

Today commanders are assigning combat roles to airmen and -women who had effectively been space watchers. One of those new roles is the Defensive Duty Officer, who must track every space launch around the world to search for new space threats. That means twenty-four hours a day looking for irregularities on a computer screen that are just as real and just as dangerous as Taliban fighters encircling a forward military outpost in Afghanistan. It was a DDO at Vandenberg's Joint Space Operations Center who first noticed Russia's kamikaze satellite in flight in May 2014.

The airmen at Schriever, Buckley, and Vandenberg are the

sentries of a looming conflict in space, there to sound the alarm when US space assets are under threat. The challenge: what to do when the alert is sounded? The men and women of the 50th Space Wing may call themselves space warriors but for now they are unarmed. And so, despite the lofty motto, the squadron headquarters is not much more than an observation post.

For now, if they see a threat and run it up the chain of command to General Buck, all he can do is ask his team in Colorado to move satellites to assess threats or get out of the way of them.

"I'll tell you this. There's not a single segment of our space architecture that isn't at risk, to include our ground segments," General Buck explained. "Satellites were built fifteen years ago. That means they were designed twenty years ago. So they were designed and built and launched during an era when space was a benign environment. There was no threat. Can you imagine building a refueler aircraft or a jet, for that matter, with no inherent defensive capabilities? So, our satellites are at risk, and our ground infrastructure is at risk. And we're working hard to make sure that we can protect and defend them."

The most pressing question on this front of the Shadow War remains open: Will the United States answer force with force? Will the United States deploy its own space weapons? We visited the place where that question may be answered. "Protecting and defending" US space assets is the most secretive element of the US space program—one that America's space commanders normally speak about only in the most oblique terms. And that highly classified strategic effort is being led, fittingly, in one of the most secretive US military installations.

US Strategic Command is located in a 1950s-era bunker buried three stories underground at Offutt Air Force Base in Nebraska. Reaching the operations floor takes you down three flights of stairs and, it seems, back to another era in time. Built during the height of the Cold War, the bunker was designed to withstand a nuclear blast. The walls are lined with thick, reinforced concrete. A series of thick steel doors lined with copper block the corridors—the steel to resist the concussive force of a nuclear blast; the copper to block the accompanying electromagnetic pulse.

Today's nuclear weapons, however, are far too powerful and too accurate for three stories of dirt and concrete to endure. So the crews here do not consider Strategic Command as a safe haven from nuclear war. A bunker would have to be hundreds of feet underground, perhaps more, to stand a chance. And so, today, the servicemen and -women based here have a failsafe. On the tarmac above sits a military jet fueled up and ready to go twenty-four hours a day, seven days a week. In the event of nuclear attack, they would escape to the air, where commanders could then continue to conduct nuclear warfare at altitude.

Throughout its history spanning more than a hundred years, Offutt has been adapting to the threat of the times. It was first founded in the 1890s as Fort Crook, a forward outpost in the Indian wars on the Great Plains. Aviation first came to Offutt in 1918, when it was home to an Army Air Service balloon field. It was later renamed in honor of Omaha native First Lieutenant Jarvis Offutt, who flew biplanes during World War *One*. During World War II, Offutt Air Force Base built the first two bombers to drop atomic bombs, the *Enola Gay* over Hiroshima and the *Bock's Car* over Nagasaki. In the 1950s, as the Cold War deepened, Offutt found itself at the center of a looming nuclear conflict as the home of Strategic Air Command, command and control for the country's most devastating weapons: its nuclear ICBMs and long-range strategic bombers.[7]

Even today its nuclear mission remains front and center. Next to a line of digital clocks bearing times from across the world—

Honolulu, Washington, Greenwich Mean Time or Zulu, Seoul, Tokyo—are three clocks reading "Red Impact," "Blue Impact," and "Safe Escape." These are the clocks that would start counting down when nuclear missiles launch: "Red Impact" for the time until US ICBMs hit an adversary's soil; "Blue Impact" for the time until enemy ICBMs impact US soil; and "Safe Escape" for the time commanders here have to escape to their airborne command aircraft.

Staring at the wall, I felt transported to 1970s and '80s Hollywood. This felt like a movie set for *WarGames* or *Doctor Strangelove*. The clocks seem borrowed from another era, before the fall of the Soviet Union, when global nuclear war was an international fixation. But the service members I spoke to make clear that the threat may have dissipated but it has not disappeared.

However, today Strategic Command is responsible for far more than nuclear warfare. It is US headquarters for a total of nine crucial and disparate missions, ranging from nuclear strike to cyber warfare, information warfare, and ISR: intelligence, surveillance, and reconnaissance. In the early 2000s it took over control of US space forces as well, when Strategic Air Command was redesignated Strategic Command. One sign of just how central Offutt is to US defense is that when the planes struck the towers on 9/11, Offutt is where Air Force One took the president. Staff here still remember the somber look on President George W. Bush's face as he walked into the operations center that morning, first informed on these grounds that a plane had struck the Pentagon as well as the World Trade Center.

Until 2016, Admiral Cecil Haney led Strategic Command. He is an example of the modern military officer: extremely well educated, with two master's degrees in engineering and technology and another in national security strategy, as well as a wealth of warfighting experience. Admiral Haney is a submariner, with multiple assignments on nuclear attack and ballistic missile submarines, one leg of the nuclear triad. After visiting virtually every unit of US Space Command, I had one overriding question for him: at a time when Russia and China are testing and deploying missiles, lasers, and kamikaze satellites capable of taking down virtually every US space asset in every orbit, how can the United States credibly defend itself without the ability to shoot back?

For a commander discussing the military's most classified programs, his answer was surprisingly clear.

"We are developing capabilities across the full spectrum. No option is off the table," he said.

And, so, thirty feet underground, in perhaps the world's most powerful command-and-control center, I learned that the first space arms race could soon be under way.

While possible or even probable, a military engagement in space would be costly for all involved. The aftermath of a wide-ranging space war would be devastating and irreversible. Hundreds of thousands, perhaps millions of pieces of debris would render earth orbits off-limits to manned and unmanned space flight for generations. Space would become an orbiting minefield with each piece of debris a potential satellite or spacecraft kill vehicle. If those busy earth orbits are no-go zones, people down on earth lose multiple capabilities they've come to depend on: satellite television, satellite communications, financial transactions timed via GPS satellites, and—perhaps most crucially—the confidence that the US military can adequately defend the homeland.

To describe the stakes of the first space war, General Hyten recalls the bloodiest battle in US military history, Gettysburg. The comparison jolted me. Memories of Gettysburg are hallowed in the US military. If this is how the nation's top space commander sees the coming conflict, just what is the United States in for?

"If you've ever been to Gettysburg, it is one of the most beautiful places on the planet," General Hyten told me, looking wistful, even tearful as he went on. "It's gorgeous. I've been there a dozen times probably. I take my family there. I take friends and we walk the battlefield. And as you walk the battlefield, it's just gorgeous and then you imagine what it was like on July the third, 1863, as Pickett made his last charge, and you realize the thousands and thousands of dead bodies that were on that field, and the thousands of dead horses and the huge destruction of humanity. Just the worst scene you can imagine.

"But if you go kinetic in space, the environment that you destroy and create is there for decades and decades and centuries and centuries and the geosynchronous exists there forever," Hyten continued. "We have to be able to avoid that. It takes away the dream of exploring space. And if you look at China, and you look at Russia and you look at the United States, you look at Europe, you look at Japan, you look at Israel, they all have a fascination with exploring the heavens. Why would we want to destroy that? So, there will be conflict. I just hope if that conflict happens, it doesn't happen in a way that it destroys the environment."

There will be a conflict. Those were disturbing words to hear from the man who would soon hold one of the most powerful posts in the US military.

US military commanders make clear that a destroyed space environment would mean a war with no winners. As space commanders like to say, "satellites have no mothers," that is, the immediate casualties of a war in space would not be flesh and blood. That, however, might make war in space more, rather than less, thinkable. A mutual assured destruction—"MAD," as the threat of nuclear annihilation was known during the Cold War—for the space age.

"I would like to think that we are smarter than that about what you could do to space if you decide that you're going to have it a free-fire zone for kinetic weapons," says General Shelton. "It would be unusable for all, not just the two [obvious] adversaries, but for everyone. And it would be unusable for a long time."

However, the nuclear standoff had, over time, an agreed-upon set of principles that allowed for negotiations, weapons reductions, and, ultimately, the avoidance of conflict.

"If not explicit, there was certainly an implicit agreement during the Cold War with the Russians," said General Shelton. "That strategic assets in space were off-limits. Both nations, at least, again, as a minimum implicitly agreed to that red line. I don't believe that we have had that same dialogue with the Chinese and therefore I don't think there's going to be near as much of a deterrent thought in their minds."

In other words, space is lawless, with no treaty to keep the peace.

The United States has tried weaponizing space before—and the immense dangers were immediately clear. In fact, the space age, inaugurated by Russia's launch of Sputnik in 1957, was just one year old when the United States introduced weapons into earth's orbit. US military commanders viewed space as the next logical battlefield.

In July 1962, the United States secretly exploded a nuclear weapon 240 miles up in space. Dubbed "Starfish Prime," the classified weapons test was the most powerful shot fired by any nation— any human being, in fact—outside the earth's atmosphere. The effects were both terrifying and mesmerizing. Observers on the ground witnessed a bright, lasting glow so strong that an aircraft far away in New Zealand found it easier to spot their targets on the surface in an antisubmarine exercise. Beyond the visible effects, the blast energized the ionosphere, an invisible blanket of electricity surrounding earth, frying electronics on the ground. The blast shut down the entire electric grid in Hawaii, though the actual explosion had taken place over a point more than a thousand miles away in the South Pacific.[8]

The US military would later discover another unexpected effect: the explosion had created man-made radiation belts around earth. Those belts of pulsing electrons would disable seven satellites over the months that followed, including the world's first commercial communications satellite and Great Britain's first military satellite. The US space program would continue to detect those radiation belts through a decade in which US astronauts took dozens of trips into earth's orbit and beyond. Despite those early warning signs, the United States and the Soviet Union would conduct several more nuclear detonations in space before the end of 1962.[9]

In 1963, Russia would conduct a further space weapons test: arming a satellite with a chemical explosive device, maneuvering it close to another Soviet satellite, and detonating it. Russia's blast had no measurable effects on earth, but it immediately littered low-earth orbit with thousands of pieces of debris, threatening its own and every other nation's satellites with the possibility of damaging or fatal collisions in space. It was a Russian kamikaze, more than fifty years before Kosmos 2499.

Those early space weapons tests happened far outside the public eye but were concerning enough for both Russian and US leaders that the Cold War adversaries declared an informal truce in space. The truce lasted—mostly—into the 1980s, when "Star Wars" again broke into the public consciousness. President Ronald Reagan had proposed his Strategic Defense Initiative (SDI), dubbed "Star Wars" by supporters and opponents alike, as a way to rid the earth of the scourge of nuclear weapons. In fact, SDI reignited a new arms race in space. In 1985, the United States conducted another space weapons test, this time fitting an F-15 fighter-bomber with a modified missile called an air-launched miniature vehicle, or ALMV, and firing it at a deactivated US weather satellite. The missile hit its target, littering low-earth orbit with thousands of pieces of debris. Russia again proposed a ban on space-based weapons. The two sides reached no formal agreement but halted tests for a time.

For the Soviets and Americans, space weapons were seen as weapons to be deployed only in a nuclear war—one part of an already catastrophic "mutual assured destruction," or MAD, scenario. There were no limited-use options in space for either side. Today both Russia and the United States, along with China, North Korea, and Iran, increasingly see space weapons as a potential part of a conventional conflict—more "thinkable" and therefore more likely to be deployed and fired in anger, creating one more front in the Shadow War.

The list of potential combatants has grown, though. Nearly a dozen countries have the capability to detonate a nuclear device in space, among them the United States, Russia, China, Iran, North Korea, as well as NATO allies Britain and France. Any country with a long-range missile system—a list even longer than that of the world's nuclear powers—can fire a missile at space targets. Meanwhile, dozens of nations are working on laser and directed-energy weapons capable of disabling or damaging space assets. Some—such as Russian-made GPS and GLONASS jammers—are available commercially, giving nonstate actors a similar capability.

———

Fifty years after those first space weapons tests, space again became a proving ground for several nations, led by the central players in the Shadow War. On January 11, 2007, China launched a rocket from its Xichang Satellite Launch Center in the mountains of Sichuan province. Some in the United States initially thought the Chinese rocket was carrying a conventional satellite. But as it rose above the atmosphere, its flight path took it on a collision course with a Chinese weather satellite. Impacting at eight kilometers per second, the rocket blasted the satellite—literally—into thousands of pieces. It had turned out to be a massive space kill vehicle. China's space weapons test sparked international condemnation, both for the ominous step toward introducing weapons into space and the more immediate danger of adding six thousand pieces of debris to the already crowded earth orbits—each piece by itself a potential satellite or spacecraft killer. One piece of space debris as small as two centimeters across traveling at orbital speeds carries the force of an SUV traveling 70 miles per hour—enough to destroy anything man has ever launched into space, the International Space Station included.

In February 2008, the USS *Lake Erie* guided missile cruiser fired a tactical missile from sea into space to intercept and destroy an orbiting US satellite that had veered out of control. Officially, "Operation Burnt Frost" was intended to prevent toxic fuel on board the satellite from threatening people on earth. But many saw a message to China, Russia, and others that the United States has some offensive space capabilities at the ready. The United States has not made a decision to deploy offensive weapons in space. However, that 2008 missile strike showed the military does at least have the capabilities, if and when the president and military commanders make such a decision to do so. And when I speak to those commanders, it is clear some are pushing to do just that.

The list of potential space weapons ranges from missiles, to directed-energy weapons, including lasers, to orbital antisatellites, such as Russia's Kosmos 2499 "kamikaze satellite," to the detonation of nuclear devices in space.

A CNN team witnessed the US military's first operational laser weapon, known as the "LaWS," or Laser Weapons System, during a test in the Persian Gulf in 2017. As the CNN team watched, an instantaneous burst of energy destroyed targets—first on the surface, then in the air—its deadly firepower moving at the speed of light. The Navy told us the LaWS obliterates targets like a "long-distance blowtorch." Mounted on the deck of the USS *Ponce*, the LaWS was not experimental, but deployed and at the captain's disposal to counter any incoming threat.

Deploying weapons like the LaWS in space would require a major strategic shift for the United States—a change US military leaders and planners are still debating. Still, many took notice in April 2016 when Deputy Secretary of Defense Robert Work seemed to issue a new warning, vowing in a speech that the United States would not hesitate to "strike back" if attacked in space—strike back, he added, "and knock them out."

Meeting him in his office in the Pentagon later that year, I asked him if he was threatening to go to war in space if necessary.

"No, the US has no intention of starting a war in space, but I do not want our potential adversaries to think, 'Well, we're just gonna be able to swing away at these guys—and sooner or later we're gonna get a lucky punch,'" Secretary Work explained. "'And we're gonna be able to knock out their satellites.' So I think a potential adversary should know that. From the very beginning, if someone starts going after our space constellation, we're gonna go after the capabilities that threaten us, and take actions to prevent them from doing that."

Such a change could be limited to deploying defensive capabilities in and around US satellites, the space equivalent of missile defense.

"Let me just say that having the capability to shoot the torpedo would be a good thing to have in our quiver," he continued, "so we're constantly trying to debate what is the best way to do it. In my view, just sitting and taking the punches is not the right way to go."

However, the US military could take weaponizing space a step further by deploying space weapons with offensive capabilities, similar to those Russia and China are currently deploying and testing.

"There's two types of deterrents, essentially. There's deterrence by denial, where you try to convince your adversary that no matter what they do, no matter how many attacks they shoot at our satellites, we're still gonna be able to operate through those attacks. And then there's deterrent by punishment. Saying, 'If you hit me, I'm gonna hit you harder.'"

It is that second option that some are now advocating for. Weapons fired from earth is one thing. In theory, every tactical missile has the ability to take out a target in space, at least in low-earth orbit, which is only a hundred miles or so above the atmosphere. The United States used a missile fired from an F-15 flying within the earth's atmosphere to take down a target in space in the 1980s and in 2008 a missile fired from a US Navy destroyer to do so again. But Secretary Work was raising the possibility of deploying weapons into space itself, arming satellites to defend themselves against threats.

"You could have weapons that could be fired against the weapons that are coming at you," said Secretary Work, who compared them to the space equivalent of the depth charges US warships fired against enemy submarines during World War II.

"Destroyers would escort merchant ships. And they would have depth charges on them to attack submarines. So, you can imagine us doing that type of activity in space. Essentially, it would all be defensive in nature, trying to keep our satellites from being destroyed. So, some people might say, 'Well, that sounds like offensive war in space to me.' We look at this as totally defensive. We have said, first of all, as a matter of policy, we don't pursue capabilities that would produce lots of debris in space."

Looking further into the realm of Star Wars, the United States has been quietly developing the first American space drone, the X37B. Bearing a striking resemblance to the space shuttle, the drone is officially a reusable spacecraft for carrying payloads into space. Its other missions are classified, but its combination of maneuverability and proven ability to orbit for hundreds of days give it the potential for both offensive and defensive roles in space. Again, the United States insists the X37B is not a weapon. But do Russia and China believe it?

"They can conclude whatever they want to conclude," General Hyten told me. "All I can tell you is right now that is not a weapon. And that's not what it was intended for. It allows us to experiment with new technology. It allows us to bring things back, look at what happened, send it back again if we want to. It's extremely useful. So, I can tell you what it is. I can tell the world what it is. And it's not a weapon. But people can look at it and they can believe what they want to believe."

Still, General Hyten sees conflict in space as inevitable, eventually with technologies seemingly drawn from Hollywood.

"Someday there'll be X-Wing fighters. It's an extension of a conflict between human beings. Every domain we've gone into has been subject to conflict and you know, it would be nice to believe it will never happen in space, but that's a mistake. We have to assume the worst case, assume what we see is true, that people are building those capabilities to challenge us, and we will have capabilities to defeat that. And if called upon, we'll do that."

———————

So how to make space war less thinkable? Space commanders say the United States must completely rethink not only its space defenses but also the vulnerability of its current space assets to attack.

"The word we're looking for here is resilience," says author Peter Singer, who has advised the National Security Council on space threats. "Resilience is the ability to shrug off the bad thing that happens to you to get back up quickly when you've been knocked down. There's a difference between resilience and Cold War–style deterrence. You don't hit me because I'll hit you back just as hard. Resilience is deterrence by denial. You don't hit me because it's not going to work. I'm going to shake it off. We don't yet have resilience in our space capability. We don't have enough satellites."

The United States got a bitter taste of that lack of resilience in 2014, when an isolated technical glitch in the GPS system hamstrung the entire US military.

"You had literally tens of thousands of US military systems— everything from aircraft carriers down to individual Humvees— that couldn't navigate," says Singer. "They didn't know where they were and where everyone else was in the system. That was a glitch. Take that scenario, move it into war, and that's the impact of the kind of conflicts that might happen if you lose space."

Satellites are, by their nature, vulnerable, designed and built to cram as many sophisticated systems into the spacecraft as possible with the least weight—to reduce the size and cost of the rocket necessary to propel the satellite into orbit. This reality makes armoring satellites the way you would a battleship, or adding defensive capabilities like you would an aircraft, far from ideal. There are other challenges as well. Chaff, for instance, the tiny bits of metal that military aircraft fire to fool antiaircraft missiles, would create clouds of space debris that would further endanger US satellites.

"They're trying to build in resilience into the systems, which means that an adversary might be able to take out some satellites, but however we can do it, we don't want to lose the mission of that satellite," Graziani said.

The focus then is on reducing the damage from losing one or a handful of satellites. This means deploying satellites that can be replaced relatively quickly and cheaply, as well as distributing a particular mission over many satellites. Remember those four satellites that serve as the nation's nuclear early-warning system? Losing just one of them would leave the United States blind to missile launches from perhaps a quarter of the planet.

"It's a strategy for dealing with the fact that the adversaries are coming after our satellites because they realized they're vulnerable," said Graziani.

The focus on resilience does not eliminate the consideration of more traditional forms of defense. Some US satellites are now being equipped with some "hardening" against electronic jamming. In addition, more satellites are being built with the capability to move out of the path of "kamikaze" or "kidnapper" satellites. These are not foolproof steps. The most powerful directed-energy weapons can defeat hardening. Fire enough energy at a satellite and its systems are fried regardless. And while some potential target satellites can move, so can Russia's kamikaze satellites and China's kidnappers.

One path to resilience may lie in the rapidly advancing technology of microsatellites—satellites as small as several centimeters across. The United States has already deployed hundreds of them in an experimental capacity, launching a dozen or more at a time, piggybacked onto other space missions. Smaller than a toaster, they can relay signals, take pictures, and maneuver through space.

"Most of it is in research and development right now," explained General David Buck of Air Force Space Command, as he handed me a microsatellite currently in testing. The black metal cube, weighing a few pounds, resembled a component from a stereo system. It looked nothing like a spacecraft. But he assured me that, as we spoke, dozens were floating over our heads more than a hundred miles up.

"The US military," he said, "is trying to find out how can we really use this miniaturized technology. Ship tracking, communication relays—some of them even have some propulsion on board."

The United States is already tracking more than two hundred operational microsatellites in orbit, launched by the United States and several other countries, including Russia and China. For the United States, the potential application is simple: spread America's space needs from the hundreds deployed today to thousands or more—create too many targets for American adversaries to take down.

For now, microsats are largely experimental. And, for the most essential missions, the United States must depend on their much larger cousins. The laws of physics are such that satellites need to be of a certain size and power to transmit signals with the energy needed to cover the large distances from space to earth and back. The United States is, however, equipping new satellites with new capabilities: thrusters and fuel to move out of the way of potential threats, shutters to block laser weapons, and perhaps soon, those "space depth charges" Secretary Work proposed.

Americans remember the space race of the 1950s and '60s as one of America's greatest victories. But they may also forget it began with the panic following Russia's 1957 launch of Sputnik. Man's first man-made satellite ignited fears that the United States was falling behind its superpower adversary. The United States soon stood up a massive space program from the Mercury astronauts, to JFK's vow to send a man to the moon, to the ultimate victory of the Apollo program with the moon landing on July 20, 1969. But this new space race—and this new front in the Shadow War—is moving quickly. And the United States faces not one, but two, capable adversaries in Russia and China.

"We're going to see a landing on the dark side of the moon, something that no one has ever done before," said Peter Singer prior to China's successful moon landing in January 2019. "A rocket is going to take off from planet earth, go to the moon, drop a lander. A lander will land on the dark side of the moon, a robot will roll out, and it will have the Chinese flag written on the side of it. It will be a historic moment, not just for China but for humanity, and we have to wrap our heads around the fact that we're not going to be the ones leading the way."

It is the arms race of the twenty-first century and beyond—with an emerging American battle plan but no certain victory. That, in itself, is an adjustment. The United States is used to winning in space. But like the space race of the 1960s, today's space race is extremely competitive.

"There's more recognition of the problem but really no more action to address the actual remedies needed," Singer said.

Secretary Work issued a call to arms at the Space Symposium in Colorado in April 2015. The US military is now fighting to respond.

I n early 2016, Paul Graziani and his team at AGI noticed that their old friend Kosmos 2499 was active again for the first time in nearly a year.

"It went quiescent for a long time, and everybody thought, 'Okay, it's out of fuel,'" Graziani told me, "because that's probably what would have happened given that you're maneuvering around so much."

Since its launch in 2014, the Russian satellite had been busy performing complicated maneuvers near and around other Russian space assets in apparent tests of its maneuvering capability. US Space Command had determined that Kosmos 2499 had the capability to maneuver into the path of another satellite to destroy it, dubbing the Russian spacecraft a "kamikaze" satellite. Russia was, the United States believed, testing a new space weapon. But in 2015, Kosmos 2499 had gone quiet, leading the United States to believe it had come to the end of its lifetime.

"Then, surprisingly, after not maneuvering for many months, it came back alive again and did some pretty serious maneuvers again," said Graziani.

The maneuvers were similar to those the United States had observed early in its lifetime; however, they were taking place after a long period of dormancy. This indicated the Russian satellite had more fuel, and more staying power, than previously known. Kosmos 2499 was in action again.

"The systems that they have launched they've continued to use and use in different ways," said Graziani.

The following year, on June 23, 2017, just over three years after the US military detected the first Russian kamikaze satellite prowling through orbit, Russia launched a rocket from its Plesetsk Cosmodrome, on the edge of the Russian Arctic. Neither the Russian Defense Ministry nor Russian state media revealed any details about the rocket's payload. Some Russian space monitoring websites theorized it might be carrying a new, so-called geodetic satellite designed to gather precision measurements of the earth's surface for use in missile targeting. However, for the first two months of its life orbiting four hundred miles above earth, Kosmos 2519 remained a mystery.

That changed on August 23, 2017, when Kosmos 2519 seemed to give birth to a smaller sibling, dubbed Kosmos 2521. However, unlike in 2015, when US military technicians had to make their own judgments about the satellite's function and capabilities, this time Moscow was uncharacteristically straightforward. The Russian Ministry of Defense announced that Kosmos 2521 was a so-called inspector satellite, specifying that it would approach and inspect its host satellite, Kosmos 2519.[10] Like China, Russia was claiming that these highly maneuverable satellites were simply space-bound repairmen, rather than weapons.

Another two months later, on October 30, Kosmos 2521 released its own smaller sibling, Kosmos 2523, into orbit—another "inspector satellite," explained the Russian Defense Ministry. One rocket launch had propelled three satellites into orbit, each with remarkable maneuvering capability. Following standard procedure, Russia registered each new satellite with the United Nations, listing the dates of their separation from their "mother" satellite as their launch dates, with the note: "Intended for assignments on behalf of the Ministry of Defense of the Russian Federation."[11]

China went into action again as well. In early 2018, SJ-17, which China had launched in November 2016, was demonstrating maneuvering capability in a new, far more distant corner of space: geosynchronous orbit, some 22,000 miles up.

It was a complicated dance in the farthest satellite orbit from earth, where some of America's most sensitive satellites are positioned, including the military's nuclear early-warning satellites. For these tests, China had positioned another, older satellite, with the designation "ChinaSat 20," in a graveyard orbit some two hundred kilometers above geosynchronous. The graveyard orbit is where nations "park" satellites that have come to the end of their lifetimes, to keep them a safe distance from operational satellites.

China then deployed SJ-17, the kidnapper satellite, to an orbit just below geosynchronous. The kidnapper stalked its prey all the way around the earth. AGI's team of technicians was watching in real time.

"SJ-17 went from below geosynchronous up to above geosynchronous to rendezvous with the satellite that was now moving [relative to its position]," said Graziani. "That was kind of a whole new maneuver that really allowed them to exercise some new and somewhat more sophisticated maneuvers to do the rendezvous proximity operations."

Moving between orbits was already complicated but it was how close SJ-17 maneuvered to its target that truly mesmerized Graziani and his team.

"It got to within a few hundred meters, which is very, very close," he said. "We haven't seen it get any closer to anything else."

To move within such proximity in an orbit so far from earth demonstrated enormous space awareness. China was demonstrating greater capability to track objects in space than the United States had been aware. The maneuver also showed the confidence China now had in perhaps the world's most advanced space weapon.

"They knew where both those spacecraft were that accurately and that they were that confident in their ability to maneuver around in that close of quarters," Graziani explained to me. "If not, then they would run the risk of running into that [other satellite]. That SJ-17 is very expensive and I really doubt they wanted to put that thing at risk.

"You had a pretty good idea of what this thing might do," he concluded.

China had proved it could capture an adversary's satellites in space.

In each case, Russia and China were refining capabilities that required satellites purpose-built for the task of, with Kosmos 2499, ramming another satellite, or with SJ-17, stealing one.

"They require lots of fuel and big engines that could get great distances—the distance to get you from one orbit to quite a different orbit," Graziani said. "That's what you really want in a weapons system so that you could jump on a target unsuspectingly from far away."

The Shadow War was expanding in space.

As China and Russia demonstrated the reliability of their new space weapons, the United States watched with alarm. Chinese and Russian progress further fueled discussions in the space community of what the United States must do now to protect its space assets. The next step is crucial. There is broad agreement that a failure to act would leave the United States in danger of losing a war in space.

"The rest of the world is starting to pay attention to this quite a bit, especially those who feel that they're threatened by either the Chinese or the Russians," said Graziani. "If you're worried about either of those, and you've got satellites, then you're worried about them coming after your satellites."

The most difficult question is whether the United States should respond to Russian and Chinese weapons with space weapons of its own. Past US presidents and senior military commanders have hesitated to weaponize space, fearing a new space arms race. As military commanders emphasize, no one wins a shooting war in space. Space orbits littered with the debris of battle would be unusable for everyone. Despite these considerations, Paul Graziani sees the Trump administration as more willing to deploy offensive weapons.

"I think that the last administration was much more sensitive about that last point. They really didn't like the O-word [*offensive*] being mentioned," he told me. "This administration, I'm guessing, isn't."

Some in the space community, including Graziani, welcome this openness. Their aim is not a shooting war in space but deterrence, which, as with nuclear war, would be founded in part on the principle of "mutual assured destruction." If the United States does take that step, even in a limited fashion, Graziani believes US leaders would be obligated to make such an offensive space capability clear to US adversaries.

"If you're going to have an effective deterrence, your adversaries need to know what's going to happen to them," he said. "It makes no sense to have a secret capability where you really mostly want to deter."

Kosmos 2499's launch in 2014 would prove not to have been an outlier or isolated experiment in space, but rather the first in a series of ongoing Russian tests and deployments of highly advanced space weapons. China would increasingly follow suit with space weapons of its own, some with unique capabilities. Kamikaze satellites, kidnapper satellites, laser weapons in space and on the ground—the newest space weapons of the newest space war. And the question remained for the United States: Would it join this space arms race? Or had it already?

LESSONS

Today China and Russia can paralyze the United States from space, disabling the most powerful military in the world and bringing America's civilian population to a standstill. Both Beijing and Moscow have tested and deployed weapons capable of depriving the United States of a whole host of technologies the public and private sector depend on. In this sense, America's unparalleled advantage in space-based assets and technologies has generated an unparalleled vulnerability—which Russia and China are seeking to exploit to grave effect.

US military planners and strategists have been aware of and focused on Chinese and Russian cyber capabilities for some time, but their awareness of and focus on those countries' offensive capabilities in space is more recent. As a result, even veterans of the US Space Command acknowledge the United States has not adequately addressed the danger and therefore risks falling behind. Only now is the US military developing a strategy to reduce and deter the threat to US space assets. But many questions, including whether the United States should test and deploy its own offensive space weapons, remain unresolved. One area of agreement is that the US military and public and private sectors must increase resilience in space, that is, spread space-based capabilities such as GPS and critical communications among more satellites to reduce the damage from the loss of one or a few of them. However, the United States is still faced with larger, strategic decisions similar to those it faced with the advent of nuclear weapons, including a central one: focus on deterrence or join a space arms race? Inherent in the challenge is the danger of escalating the standoff in space to a point at which the United States and its adversaries find themselves in a conflict neither side wants.

CHAPTER 7

Hacking an Election

(RUSSIA)

Russia fired the first warning shot in a bold information war on the US political system in 2014, a full year before its first probing attacks on the Democratic Party in the 2016 presidential election. The target was the email system of the US State Department. Monitoring and observing them throughout was Rick Ledgett, then deputy director of the National Security Agency (NSA) and recently departed head of its Threat Operations Center. He had been tracking and responding to Russian cyberattacks for years at the NSA, but this time, he noticed something different.

"For a very long time, when we would find Russians, and engage them in the network, do things in the network that would indicate to them that we knew they were there," Ledgett explained, "they would take defensive actions. They remove malware, things like that. They would disappear.

"They go deep, and they'd come back dressed completely differently, and you'd have to redetect them again," he added.

These had become the rules of a game of cyber cat-and-mouse—and Russia's tactics were relatively conservative and predictable. If the NSA detected Russian state hackers gaining entry, or attempting to gain entry, inside a US government network, the United States would take defensive actions to block their entry and kick them out. Russia would then abandon the network, to return another day via a different pathway and under a different cover.

"They would go home, and they would change their tools," Ledgett said. "They would change the way they looked, so that we wouldn't recognize them next time. Their principal goal was not to get caught."

That changed with its attack on the State Department in 2014. Now, when NSA technicians identified and engaged their Russian adversaries, the Russian hackers didn't leave. They would simply deploy new iterations of the same cyber tools and attack the network again. Russian hackers abandoned subtlety for blunt force.

"Beginning in 2014, their principal goal was get the data," Ledgett said. "'We don't care if you know we're here.'"

Russia's hack of State Department systems began, as most cyber intrusions do, with Russian hackers identifying and exploiting a weak link. The weak link in this case was the State Department's unclassified email system. The State Department operates two separate email networks: a classified system, known by State employees as the "high side," and an unclassified system, referred to as the "low side." State Department rules require that classified or sensitive information is shared only on the "high side." The "low side" network is also considered secure, but few foreign service officers treat it that way. When I was working at the US embassy in Beijing, my colleagues and I knew that China had access to our low-side emails. Virtually everyone had stories of Chinese diplomats suspiciously calling immediately after we had sent emails with questions apparently based on information gleaned from those communications.

While the unclassified network is not used for classified information, it still contains a wealth of information of value to foreign adversaries.

"There is a lot of interesting information that happens on the unclassified network," said Ledgett. "Indicators, pieces of information that you could put together, and draw an interesting intel picture. Second, you always look for a connection between the unclassified network and the classified network. Maybe there is one. Maybe there's not. There's usually not supposed to be, but sometimes there is."

This is the grunt work of intelligence gathering, made easier by the vast capabilities of computing in the digital age. Today intelligence agencies are in the business of big data: gathering a mind-boggling amount of emails, phone calls, calendar entries, Web searches, and more, and building them into a picture of their adversaries' activities.

"If you think about the State Department, it's a global organization," Ledgett explained. "There are unclassified terminals all around the world. They're used for everything from sending unclassified State Department cables back and forth, to ordering food for the office party."

In some countries, the State Department system was the only access to the internet. There is no Comcast in Uzbekistan, noted Ledgett as an illustrative example. And so US diplomatic staff often access the low side for personal use. The State Department sets rules for what its overseas employees can and cannot do on those systems. But those rules are often not enforced.

Russia's path into the State Department system began in one computer in one of those countries. While the exact circumstances of the State Department breach remain classified, to illustrate, Ledgett used the example of an ambassador who lets his son use the State Department system to play a Web-linked video game: one game, one opening to Russian malware, and the entire system could be exposed, giving hackers access to the activities of thousands of US diplomats in 190 countries around the world.

Russian hackers would roam inside the State Department's email system for months before the intrusion was discovered. However, Ledgett says the NSA had no doubt about who the hackers were working for.

"There's things you can look for, when you're looking for a cyber intrusion. There's the code they use, the infrastructure that they use, in other words, the 'hop points' that they use to get to where they're going," he said. "They'll use this particular server, or this particular computer, somewhere in the world, as a path to get to work from the target network. What you'll see is a sequence of operations."

Hackers also leave behind digital fingerprints that allow the NSA, over time, to recognize and identify particular hackers and their work. Like graffiti artists or poker players, hackers have "tells" that show up consistently in their cyber activity.

"It could be that they reuse code fragments," he said. "I wrote this code that's a really elegant piece of code, so to accomplish that same thing in another piece of malware, I'm going to reuse it."

That combination of laziness and arrogance is extremely useful to the NSA in determining "attribution" for this and other Russian hacks. However, Ledgett emphasizes that assigning attribution with confidence, as the NSA did with the hack of State Department emails and, later, the 2016 election, requires a mosaic of information and intelligence gathered over time.

"There is no one thing that says, 'Oh, that's this entity or that entity.' Then you also look at what they are going after," he said. "If you see someone in the network, they're stealing information, well, what are they stealing? There are things, over time you can look at that and align with, oh okay, they're stealing information on US policy towards Russia."

Clues gathered and fingerprints identified by the NSA in 2014 would become useful in determining the origin of a more expansive Russian cyberattack on the 2016 election still to come. Still, today, with clear frustration, Ledgett looks to 2014 as one more missed warning sign of a marked and dangerous change in Russian behavior. In fact, he describes consistent and repeated warning signs that Russia was qualitatively altering its cyber tactics and goals— followed by consistent and repeated underestimation by US leaders and policy makers, including in the intelligence community.

The US response to hacking followed a similar pattern as its response to other Russian attacks: missed warning signs followed by penalties insufficient to deter future attacks. The events of 2014 were particularly egregious because Russia was escalating its attacks and interference on multiple fronts at the same time: in cyberspace against the United States and on the ground against Ukraine. The escalation of Russian cyber tactics in 2014 mirrored Russian aggression around the world.

"It directly parallels what was going on in the political realm, and in the physical, kinetic realm, with the invasion of Ukraine, and in the diplomatic realm, with Russia being aggressive," said Ledgett. "And Putin's desire to regain some of Russia's relative status in the world, to the United States."

Russia was perpetrating a dizzying array of aggressive actions targeting the United States. In one particularly startling twenty-four-hour period in July 2014, a Russian missile took down the

Malaysian passenger airliner MH17 over Eastern Ukraine, while, the next day, Russian military radar locked on to a US surveillance aircraft in international airspace over northern Europe. The US flight crew was so concerned it was being targeted for attack that it fled without warning into Swedish airspace.[1] According to Ledgett, these activities all pointed to a coherent, whole-of-Russian-government direction to be more aggressive against the West.

Soon Russia would launch its boldest ever cyberattack against the United States.

General James Clapper—as director of national intelligence under President Obama, the nation's seniormost intelligence official—says he first learned Russia was attempting to infiltrate US political organizations in the summer of 2015, months after the State Department email system was compromised and more than a year before Election Day 2016. However, General Clapper admits it was not immediately clear to him how serious Russia's efforts were.

"I don't think it was, because obviously the Russians have us as a primary, maybe the primary intelligence target anyway," Clapper told me.

The specific target this time was the Democratic National Committee (DNC). The DNC received the first quiet warning in September 2015, when a midlevel FBI agent called to notify the committee that Russian hackers had compromised at least one of its computers.

Years later, Democratic Party officials still recall the FBI's initial response with thinly veiled anger.

"They left a phone message at the help desk of the DNC," said John Podesta, then chairman of Hillary Clinton's presidential campaign. "They didn't treat it with the kind of seriousness, I think, that it deserved."

As I sat across from Podesta, his frustration was palpable. He described the attack and his involvement almost in the way someone speaks about a lost loved one.

The September phone call was the FBI's first direct contact with the DNC—a message left for a low-level computer technician on the equivalent of a corporate computer help line. The technician did not return the FBI's call.

"You know, the bureau—it's a busy place," said Steve Hall, the former CIA station chief in Moscow. "They have got lots of stuff to do, but I suspect if they had to do it over again, they probably would try to do it differently in retrospect."

The DNC technician did scan the system networks but found nothing and did not share the FBI's concerns with any of his superiors at the committee. In fact, the breach would prove to be enormous. The hackers over the coming weeks and months would gain access to emails and documents numbering in the hundreds of thousands. Hall believes the Russian hackers' success likely surprised even the hackers themselves.

"I can imagine seeing them a couple of weeks later and saying, 'Well, that went pretty well. Look, we're in,'" said Hall.

For weeks, the FBI kept calling the same computer help desk number at the DNC. DNC officials complain that agents from the bureau never made the short trip across the National Mall from FBI headquarters to the headquarters of the DNC to warn them in person.

"Looking back, I think they would probably say, 'Geez, we should have been a little bit more aggressive,'" says Hall. "But again, it's hard to predict where these things are going to end up and this one ended up in a pretty interesting place."

That same lack of urgency would plague multiple agencies and political organizations as Russia's hacking expanded in the months leading up to the election—even as the attacks grew bolder.

In November 2015, one year before Election Day, the same FBI agent called the DNC once again with even more alarming news: a DNC computer was now transmitting information back to Russia. Once again, the DNC computer technician took no action—and the DNC says the FBI made no concerted effort to alert more senior members of the committee's leadership. Those decisions allowed the Russian hackers to roam freely inside DNC computers for months more, allowing them to vacuum up more information to be released later to enormous effect.

"A hostile foreign power is trying to actively engage in our electoral process, you would have thought that that would have risen up to the attention of the other intelligence agencies, of the White House itself," Podesta said.

Tom Donilon was President Obama's national security advisor until 2013. He had met Vladimir Putin face-to-face before. And, in retrospect at least, he recognized the former KGB agent's hand.

"It was alarming because it was absolutely consistent with Putin's intent to undermine the institutions of the West," Donilon told me.

"There is no doubt in my mind that Vladimir Putin was involved from the very beginning, knew all of the details of it and indeed, might have been the intellectual author of quite a bit of it," said Donilon. "He knew about this and was probably very eager to see. 'Geez, are we really going to be able to pull this off.'"

By now, Russian hackers had been roaming through the networks and servers of the Democratic National Committee for months. But they were setting their sights on new political targets. To draw in new prey, they would employ the crudest of cyber weapons: so-called spear-phishing emails. It's likely most people reading this book have been on the receiving end of similar probing contacts.

"In addition to the organizations that were targeted, multiple individuals were targeted with spear-phishing emails that resembled Google warnings," said John Hultquist, the director for intelligence analysis at the cybersecurity firm FireEye, which was later brought in by the Democratic Party to diagnose and address the hacking.

"They clicked on those thinking that they were security warnings and those basically transported them to a place where the adversary could collect their credentials and reuse them to gain access to their accounts," Hultquist explained. "It was a reasonably realistic email. It looked fairly legitimate."

Spear-phishing emails are the bedbugs of information systems: ubiquitous and almost impossible to eradicate. No matter how robust a network's firewalls or how sharp its cybersecurity team, one reflexive click by one careless employee and the entire system is vulnerable to infestation by hackers. In this case, the weak link would be the email account of Clinton campaign chairman John Podesta.

"There was a Google alert that there was some compromise in the system and that I had to change the password," Podesta, then chairman of Hillary Clinton's presidential campaign, told me in a low voice. It seemed to me he had recounted that moment in his head more than once, almost to the point of physical pain.

This seemingly benign message was actually a spear-phishing email. It warned in language common to such emails: "Someone just used your password." It then prompted the recipient to change their password "immediately." It was signed—innocuously and apparently convincingly—"Best, The Gmail Team."

"It actually got managed by my assistant who checked with our cybersecurity guy," Podesta recounted. "And through a comedy of errors, I guess, he instructed her to go ahead and click on it and she did."

That comedy of errors culminated in this fatal one: Podesta's cybersecurity guy correctly identified the email as illegitimate, but he wrote back with what would turn out to be a typo for the history books.

"He meant to say that it was illegitimate, [but] he said it was legitimate," said Podesta with utter resignation.

"The rest is history," he added.

One typo and one click, and Russian hackers had taken another step inside the Democratic Party—gaining free rein through tens of thousands of emails to and from the man running the campaign of the presumed front-runner for US president. And Russia had achieved all this with the simplest of cyber weapons.

"That is one of the frustrations I think for all cybersecurity experts," General Clapper told me with his characteristically dry delivery.

Eight months to Election Day, Russian hackers had now successfully breached two Democratic Party computer systems: that of the Clinton campaign and the Democratic National Committee. And just as the NSA's Ledgett had noticed with the hack of State Department emails a year earlier, they were not aggressively or even minimally attempting to conceal their activity.

"We were kind of blown away by the brashness that they are operating with in the space, almost like they didn't anticipate any consequences for their actions," said Hultquist.

On the campaign trail, Hillary Clinton was looking more and more like a lock for the Democratic Party's nomination. And she began to turn her attention and her attacks to her likely Republican opponent, Donald Trump. In April 2016, she revealed what would become her bumper sticker campaign slogan: "Love trumps hate." Trump was focusing his own attacks right back at Clinton, proclaiming that same day in April, "We are going to beat crooked Hillary so badly that your heads will spin."

Back at the DNC, it had now been nine months since the first intrusion by Russian hackers into the DNC's computers, when that DNC computer technician finally discovered the breach. The DNC notified the FBI and hired the cybersecurity firm Crowd-Strike.

CrowdStrike's technicians went to work and quickly identified two likely culprits, both with a long history in cyberattacks and with numerous links to Russia. Dubbed "Fancy Bear" and "Cozy Bear," US intelligence believed they were covers for hackers operating on behalf of the Russian government.

The actors were well-known to cybersecurity experts—and not particularly secretive, either. Again, their boldness struck many as unusual, even clumsy.

"We've known these actors long before—long before any of this transpired around the election," said Hultquist. "We've known these actors for many years. There's a lot of evidence that this actor is Russian or a Russian speaker."

The evidence was surprisingly simple, even to laymen. For one, the hackers appeared to be starting and finishing their workdays on Moscow time.

"The mistake they've made is leaving these time stamps," Hultquist explained. "And if you look at enough of them over time, you get a picture of what actual hours this operator is working. And what they come down to is a work schedule that fits right in with Western Russia's time zone."

The hackers left other clues as well, tying them even more directly to Vladimir Putin's Russia. The clues were what cybersecurity experts call "language artifacts." That is, they were writing computer code in the Cyrillic, or Russian, alphabet.

Hultquist and his team did not let their guard down for new intrusions. And in the summer of 2016, they detected "Fancy Bear" sniffing out new prey.

"It's really exciting to catch these guys in the act," said Hultquist with an almost mischievous excitement.

Hultquist and his team caught the hackers inside another Democratic Party entity: ActBlue, a fund-raising website for party and other progressive groups.

"They were diverting people who went to the ActBlue system, the donation system, to a server that they owned," he said.

At FireEye's headquarters, Ben Read, the firm's manager of cyber-espionage analysis, later demonstrated how the hackers replaced the genuine site with a nearly identical fake one. On his computer screen, he called up an image showing exactly how the website of the Democratic Congressional Campaign Committee, the DCCC, appeared on July 19, 2016. The link to ActBlue was indistinguishable unless someone clicked on the source of the Web page, that is, the computer code behind what appears on the screen. On the real site, the hyperlink would take the user to "secureactblue.com." On the hacked page, however, the link directed users to "secure.actblues.com," the same, it seemed, except for that extra dot and that extra *s*.

Read's immediate reaction at the time was "That was weird."

"Actblues.com," with that telltale *s*, had nothing to do with the DCCC. It was a Russian cover—and one that a minimally competent technician could find, if he or she were looking for it.

"The second I saw it, we were sitting, you know, there are emails flying everywhere, including to the target organization," said Read. "We obviously want to give them a heads-up."

This intrusion was identified and corrected before the hackers could go further into the DCCC's fund-raising system and its wealth of user and campaign data. However, it was one more attempted cyberattack by a Russian adversary seemingly uninterested in covering its tracks.

"We have high confidence that this is a Russian intelligence organization," said Hultquist. "Because we've been tracking this actor for so long and we've seen so many artifacts, forensic and otherwise, that suggest that this actor is carrying out Russian intelligence missions."

The question now was what would Russian intelligence do with all this information? Having stolen it, were they going to weaponize it?

In June 2016, five months before Election Day, the American public got the first hints of what was to come. The messenger was the mysterious blogger(s) nicknamed "Guccifer 2.0." Like Cozy Bear and Fancy Bear, US intelligence believed Guccifer was a cover for a sophisticated Russian hacking operation.

"They love putting on these false personas and carrying out operations through them," said Steve Hall, the former CIA Moscow station chief.

Most important, Russia was mobilizing the next step in their information operation against the US election: weaponizing the stolen data to influence the electoral process. Guccifer released a sampling of documents stolen from the DNC, including a list of million-dollar donors and an opposition research report on Donald Trump.

"It is oftentimes interesting to put myself, given my background, into the position of the Russian intelligence officer," said Hall. "And so, I can imagine these guys saying, 'Would it be too crazy to think that we could perhaps actually pull together an influence operation?'

"And you know, this probably ran up the chain and somebody said, 'Well, you know, why not? Let's give it a shot,'" he said.

Guccifer released the materials one day after it was first reported that the DNC had been hacked. It would go on to release stolen data not just from the DNC but also from the Clinton campaign and the DCCC, all of which had now been infiltrated by Russian hackers.

"It's the Russian services taking that gold that they have collected, making something out of it, that in turn can be used against their adversaries and in this case, the United States," said Hall.

"And we continue to be their primary enemy," he added.

Guccifer would soon be joined by another rogue publisher, led by WikiLeaks, the organization founded by Julian Assange. I had interviewed Assange in London in December 2010, soon after WikiLeaks had leaked thousands of stolen US diplomatic cables from around the world. At the time, the leak was the largest ever of classified materials. In our conversation, he explained his commitment to exposing US government institutions as, in his view, entities of an authoritarian government.

"Security officers have a job to keep things secret. The press has a job to expose the public to the truth," he told me. "So that is our job and we're doing it. The fact that the State Department was not able to do their job is a matter for them."[2]

It was a mission statement that would ring familiar six years later during the 2016 presidential election. When I pressed him why he didn't focus similar attention on actual authoritarian regimes in China and Russia, he didn't offer an answer.

On July 22, 2016, WikiLeaks posted a stunning announcement on Twitter: it would release some 19,000 emails from the Democratic National Committee.

The US intelligence community had no doubt that WikiLeaks

had obtained the emails from hackers working for the Russian government under orders from the Kremlin.

"You want what we would call in the business, a cutout," said Hall. "You want a third party. Somebody like, oh, I don't know, WikiLeaks."

The function of a cutout was to provide distance between the hacking of US political institutions and the Kremlin. Such plausible or even implausible deniability is an essential feature of the Shadow War.

General James Clapper, the former director of national intelligence, told me US intelligence agencies were not fooled. They had hard evidence of a connection between WikiLeaks and Russia.

"We were pretty high in our confidence that's what happened," Clapper told me. "I'll put it that way."

The Democratic Party convention was fast approaching. At the time, opinion polls gave Hillary Clinton a comfortable lead over Donald Trump in the general election. The convention was an opportunity not just to begin focusing her campaign's attention on her Republican opponent, but to make peace with Democratic voters who had supported Bernie Sanders for her own party's nomination.

Just three days before the convention, WikiLeaks made its first batch of emails public. The stolen emails suggested that senior leaders of the Democratic National Committee had been biased in favor of Clinton over Sanders. DNC chairwoman Debbie Wasserman Schultz was the focus of Sanders supporters' ire, who made their anger clear as she took the podium.

"Okay, everybody, now settle down!" she shouted into a chorus of boos.

Wasserman Schultz was forced to resign as chairman, making her the first victim of Russia's broadening influence operation on the election.

The subsequent email releases quickly became lead stories for US media, a phenomenon that would later fuel criticism from Clinton campaign officials and supporters that the media had too easily played foil to a foreign influence operation.

Donald Trump took particular public satisfaction in the fissures now exposed inside the Democratic Party. On July 25, he tweeted, "The new joke in town is that Russia leaked the disastrous DNC e-mails, which should never have been written (stupid), because Putin likes me."

On July 27, Trump, in a now-infamous speech, took the alarming step of encouraging Russia to hack Clinton's emails, declaring, "Russia, if you're listening, I hope you're able to find the thirty thousand emails that are missing."

US intelligence officials would later tell me that public comments such as this one by the president factored into their own suspicions of possible cooperation between the Trump campaign and Russians. As one senior intelligence official remarked to me, don't miss the evidence that is "open source," that is, evidence in the public realm.

One of Trump's longtime confidants made his own public references to materials stolen by Russia. Roger Stone repeatedly hinted that he had advance knowledge of upcoming releases of emails, via his contacts with Julian Assange of WikiLeaks.

On August 8, 2016, Stone told a Republican group, "I actually have communicated with Assange. I believe the next tranche of his documents pertain to the Clinton foundation. But there's no telling what the October surprise may be."

On August 21, he tweeted, "Trust me, it will soon [be] Podesta's time in the barrel."

In early October, Stone tweeted a series of apparent warnings of upcoming releases by WikiLeaks damaging to the Clinton campaign.

On October 2: "Wednesday Hillary Clinton is done. #Wikileaks."

On October 3: "I have total confidence that @WikiLeaks and my hero Julian Assange will educate the American people soon. #LockHerUp."

On October 5: "Libs thinking Assange will stand down are wishful thinking. Payload coming #Lockthemup."

Then on October 6, "Julian Assange will deliver a devastating exposé on Hillary at a time of his choosing. I stand by my prediction."[3]

"There seems to be some indication or some contact between forces closely associated with the Trump campaign and WikiLeaks," former DNI Clapper told me, adding dryly, "it certainly seemed like an interesting coincidence, let's put it that way."

Stone would consistently deny communicating with Assange directly, or, more broadly, colluding with Russian actors.

One person who seemed to be enjoying the escalating allegations was Vladimir Putin. In an interview with Bloomberg News on September 1, Putin dismissed the idea of Russian involvement as inconsequential.

"Does it even matter who hacked this data?" Putin said. "The important thing is the content that was given to the public.

"There's no need to distract the public's attention from the essence of the problem by raising some minor issues connected with the search for who did it," he added. "But I want to tell you again, I don't know anything about it, and on a state level Russia has never done this."[4]

Inside the Office of the Director of National Intelligence, General Clapper's concerns were growing.

"I did have a very visceral feeling in the pit of my stomach that

I thought that this was a really serious thing, an assault on the very heart of our democracy," Clapper said.

Visceral, gut-wrenching reactions were ones I heard from numerous intelligence officials involved in tracking Russian interference. They were seasoned operatives who had followed Russian intelligence operations dating back to the Cold War. But the scope and brashness of Russian interference in the 2016 election was unprecedented. Many felt a patriotic duty—and fear—they had rarely experienced before.

"That's one of the reasons I felt so strongly about putting out the statement that we did in October," Clapper told me.

On October 7, one month and one day before Election Day, US intelligence agencies publicly named Russia as the culprit for both the theft of Democratic Party materials and the strategic release of those materials over time to influence the election.

"The U.S. Intelligence Community (USIC)," read the statement from the Department of Homeland Security and the Office of the Director of National Intelligence, "is confident that the Russian Government directed the recent compromises of e-mails from US persons and institutions, including from US political organizations. The recent disclosures of alleged hacked e-mails on sites like DCLeaks.com and WikiLeaks and by the Guccifer 2.0 online persona are consistent with the methods and motivations of Russian-directed efforts. These thefts and disclosures are intended to interfere with the US election process."[5]

The selection of DHS and ODNI as the two signatories of the statement was deliberate. DHS's involvement made clear that the United States viewed Russian election interference as an attack on the homeland. The involvement of ODNI, which oversees all the intelligence agencies, indicated this was the consensus view of the US intelligence community.

President Trump and some of his supporters would later claim the assessment was a minority view because not all seventeen intelligence agencies signed on. In fact, most US intel agencies had no role in the assessment of Russian interference because they had no role in assessing such threats. Among the other agencies are the US Coast Guard's intelligence branch, which focuses on threats by sea; the Drug Enforcement Administration's intelligence branch, which focuses on intelligence related to drug trafficking; and the US Marine Corps' intelligence branch, which focuses on battlefield intelligence for deployed Marine units. Regardless, the "not all seventeen agencies" narrative would remain a lasting talking point for those doubting Russia's involvement.

While the October 7 statement focused on the theft and release of stolen Democratic Party emails and documents, it made an ominous reference to something more alarming: potential attacks on actual voting systems. "Some states have also recently seen scanning and probing of their election-related systems, which in most cases originated from servers operated by a Russian company," read the assessment. At the time, ODNI and DHS did not yet have the confidence to assess that the Russian government was also behind the attacks on election infrastructure. However, the warning foreshadowed later activity by Russia in 2018 and beyond.

With the public statement by ODNI and DHS, the US government was laying down a marker for Russia as Election Day approached. Their message: we know what you're up to and we won't allow it. However, the minutes and hours that followed quickly diverted the nation's attention.

John Podesta, chair of Hillary Clinton's campaign, told me he will never forget that evening.

"The director of homeland security and the director of national intelligence released a statement that the Russians were actively interfering in the election," Podesta recalled. "Later in the day, the *Access Hollywood* tape came out."

The release of a previously unknown recording of Trump in 2005 engaged in a lengthy, lewd conversation with *Access Hollywood* commentator Billy Bush quickly drove the ODNI assessment out of the news. Trump's remarks, on the set of the soap opera *Days of Our Lives*, where he was making a guest appearance, were mesmerizing in their misogyny.

"I did try and fuck her. She was married," he said, continuing, "I moved on her like a bitch. But I couldn't get there. And she was married. Then all of a sudden, I see her, she's now got the big phony tits and everything. She's totally changed her look."

He went on, uttering the phrase that would come to define what became known simply as "the *Access Hollywood* tape."

"I'm automatically attracted to beautiful—I just start kissing them. It's like a magnet. Just kiss. I don't even wait. And when you're a star, they let you do it. You can do anything," he said. "Grab 'em by the pussy. You can do anything."

"Of course, everyone's attention turned to what Donald Trump had been saying to Billy Bush on that bus," remembers Podesta.

Inside the Clinton campaign, the reaction was both shock and celebration. Already confident in victory, Clinton campaign staffers began talking about when Trump would withdraw from the race and who might replace him. One campaign staffer told me that night that Trump was finished.

The time of the tape's release by the *Washington Post* was 4:02 p.m., Eastern Time. It appeared to be an enormous break for the Clinton campaign. But twenty-nine minutes later, WikiLeaks stunned the world with this tweet: "RELEASE: The Podesta Emails #HillaryClinton #Podesta #imWithHer" along with a link to the stolen documents.

"Within minutes, the first of the emails was posted to WikiLeaks," recalled Podesta. "With a statement from Julian Assange that said that we have the contents of his email system and we're going to release them all during the course of the campaign."

WikiLeaks revealed it had acquired the entire contents of John Podesta's private email account, totaling more than fifty thousand emails, including thousands between him and every senior member of Hillary Clinton's presidential campaign. Officials in the US intelligence community suspected the release had been timed for maximum impact. Once again, the evidence was hiding in plain sight. Members of the Clinton campaign, not surprisingly, agreed.

"Let's just put it this way," said Podesta. "It's a pretty massive coincidence that they would choose to pull the trigger on a Friday evening, when they've been sitting on it for a while."

Hillary Clinton made her concerns public immediately, telling reporters, "No, I have nothing to say about WikiLeaks other than I think we should all be concerned about what the Russians are trying to do to our election."

That Friday-night email dump would turn out to be just the first of many. WikiLeaks would trickle out the stolen emails in tranches of a thousand or so apiece, every few days right up to Election Day. And those releases became a dominant story line of the campaign.

"The Russians are pretty intense observers of what goes on in this country," Clapper told me. "And [they] tried to both collect information on it and, as we saw, where they can, exploit it."

On the campaign trail, a bizarre dynamic was emerging. While Hillary Clinton was raising the alarm about a foreign adversary's intrusion into the US electoral process, Donald Trump gleefully egged them on.

"WikiLeaks. I love WikiLeaks!" he proclaimed at a campaign event on October 10.

Three days later, he told a campaign crowd, "It's been amazing what's coming out on WikiLeaks."

Then, on October 31, "This WikiLeaks is a treasure trove!"

On November 4, four days before the election, he declared once again, "Boy, I love reading these WikiLeaks."

Then-candidate Trump and some of his supporters would dismiss Russian interference as insignificant and not particularly new. Russia had indeed been attempting to interfere in US elections for decades. However, intelligence officials who oversaw the US response to Russian meddling in 2016 say the scope and intensity were unprecedented.

"It was the weaponization of the information that they took from the DNC. That was different," former NSA deputy director Rick Ledgett told me. "Now they're going to deploy it, to try to influence the election."

Later in October, Russia's weaponization of stolen Democratic Party emails would expand to new targets in the Clinton campaign, including Neera Tanden, a longtime Clinton confidante and a member of her transition team. Tanden first learned of the breach on the news.

"I think I saw my name on like one of the TVs," Tanden told me. "And I was like, 'What happened?'"

In the stolen emails, Tanden could be seen criticizing other campaign staff and Hillary Clinton herself. In one email, Tanden wrote that whoever allowed Clinton to use a private email server should be "drawn and quartered." In another, she says that by failing to articulate a position on the Keystone XL pipeline, Clinton was "dodging another issue." In perhaps her most infamous comment, Tanden wrote, "Hillary. God. Her instincts are suboptimal."

Once again, Donald Trump reveled in the revelations. In the third and final presidential debate on October 19, Trump said, "WikiLeaks just came out. . . . Now John Podesta said you have terrible instincts. Bernie Sanders said you have bad judgment. I agree with both."

He had misquoted the emails. It was Tanden, not Podesta, who called Clinton's instincts "suboptimal," but regardless the material was white-hot ammunition for the Trump campaign, as well as a zinger of a line for Trump in a nationally televised presidential debate.

"He misjudged it, saying it was John Podesta," Tanden later told me. "It was mine. But I remember watching the TV as that happened and wanting to put my head under the pillow."

The releases of the stolen Clinton campaign emails and memos would continue right up to Election Day. While confident in public, in private some Clinton campaign staff began to worry the releases could lose her the election.

"All these emails were just getting dumped day in and day out. And every morning, I basically woke up with dread for what was going to come next," said Tanden. "And I was like, 'Is this going to sink the campaign or not?' "

Inside the White House, a sometimes bitter debate was unfolding, pitting senior advisors, including Secretary of State John Kerry, who were pushing for a more robust US response, against others, led by President Obama, who feared both escalation with Russia abroad and allegations of influencing the election at home.

In a press conference one month after the election, Obama explained his concerns. "At a time when anything that was said by me or anybody in the White House would immediately be seen through a partisan lens, I wanted to make sure that everybody understood we were playing this thing straight."[6]

As Election Day approached, the Obama administration's greatest fear was that Russia would disrupt or attempt to disrupt actual voting systems, including voting machines and voter registration databases. Russia would only need to interfere in voting in a handful of districts to call the entire election into question. Particularly in a close vote, the consequences could be disastrous.

At the summit of G20 leaders in Beijing in early September 2016, President Obama had warned President Putin face-to-face not to interfere on Election Day.

"I felt that the most effective way to ensure that that didn't happen was to talk to him directly and tell him to cut it out and there were going to be serious consequences if he didn't," the president said on December 16.

Later, President Obama made rare use of a direct messaging system between the White House and the Kremlin, originally intended to avert nuclear war, to warn Putin once again.

To this day, Clinton campaign advisors struggle to assess how much Russian interference damaged their campaign and their candidate, Hillary Clinton.

"Look, it was our job to win and we didn't do it," Podesta told me. "What went into that? You know, a lot of things. And we bear our own sense of responsibility for that. But I think it was an important element of electing Donald Trump and I think the Russians got what they paid for."

It was only after Election Day, with a new president-elect few including President Obama had expected, that the Obama administration finally retaliated more substantively. Obama ordered the closure of two Russian diplomatic compounds that the United States believed were used for spying. He expelled some thirty-five Russian diplomats, most or all of whom US intelligence believed were intelligence agents working under diplomatic cover. The administration also imposed new economic sanctions on Russian individuals and entities.

In secret, Mr. Obama considered taking more aggressive steps, including initiating an NSA plan to place cyber weapons inside the computer networks of critical Russian infrastructure. Those systems could have been activated if Russia were to carry out new cyberattacks on the United States. However, for now, US retaliation was complete as the nation awaited its new president.

Today, former NSA deputy director Rick Ledgett believes the US response to Russia's interference in the 2016 election was weak, therefore inviting future Russian cyberattacks on US elections.

"We gave them the go-ahead," Ledgett told me, his frustration impossible to miss. "If you allow bad things to continue to happen, despite the fact that you know it, you're setting a policy. You're setting a precedent. You're defining that as acceptable behavior."

Ledgett said his criticism applies more broadly to the US response to Russian cyberattacks targeting a range of critical infrastructure.

"I think we don't have a coherent way to respond to this, so they keep going," Ledgett told me. "So, what are they going to do next?"

NSA officials have stated repeatedly that the United States has unrivaled cyber capabilities, outmatching Russia and China. However, US intelligence officials concede that the United States has unmatched vulnerabilities as well, due to the country's dependence on technologies that are most vulnerable to cyberattack, from communications networks, to satellites, to power grids, to elections.

"The US is not in a good position to get into a cyber war with other people, because we are more vulnerable than almost anybody else in the world," said Ledgett. "It's the old adage about, you don't start a rock fight if you're living in a glass house."

US adversaries are keenly aware of American dependency and so are constantly looking for new ways to exploit US vulnerabilities.

Those efforts are not confined to foreign intelligence services. US officials believe that private Russian technology companies are required by law to provide Russian government agencies access to their technologies. This includes Russia's best-known international technology provider, Kaspersky Lab. Kaspersky antivirus and cybersecurity software was long ubiquitous in the United States. However, cybersecurity experts believe Kaspersky products contain so-called back doors accessible to Russian intelligence. Kaspersky has repeatedly denied the existence of such back doors.

"By law, if the Russian intelligence services tell [Kaspersky] to give them access to information they have, they're required by law to do that, regardless of where they are operating in the world," said Ledgett. "Any company operating in Russia, and any Russian company, operating anywhere in the world. That's what the law applies to."

Still, it was not until December 2017—more than two years after Russian hackers first penetrated US political organizations to interfere in the 2016 election—that Congress passed a law banning Kaspersky software from all US government computers.

Senator Jeanne Shaheen, who helped sponsor the legislation, called Kaspersky products a "grave risk" to US national security, adding, "The case against Kaspersky is well documented and deeply concerning. This law is long overdue."[7]

Ledgett and other US intelligence officials see the growth of "the internet of things," bringing a whole host of internet-connected devices into American homes, from refrigerators to voice-activated internet-connected devices such as Alexa, as a new, clear, and present danger. Ledgett, for his part, bans any instance of "the internet of things" from his own home.

"There are no real standards yet, and the devices just don't have the kind of security protocols that I would want," said Ledgett. "I don't have an Alexa, or the Amazon version of that, because it's a directional microphone that can be remotely controlled."

One prominent US vulnerability has no easy fix. Russian interference in the 2016 election showed that the country's sharply divided politics make for fertile ground for Russian influence operations. Fake news did not need to be force-fed into the American political conversation. It was swallowed whole. At times, Presidents Trump and Putin seem to be sharing talking points. One powerful example is enduring doubts about Russian interference in 2016.

Rick Ledgett seethes at the very question.

"I personally looked at every single piece of intelligence that went into the intelligence community assessment, every piece of intelligence from every part of the intelligence community," Ledgett told me. "I spent seven to eight hours, somewhere in that neighborhood, talking to our analysts, who participated in the community assessment, and I will tell you, there is absolutely no doubt that it was the Russians, and it was directed by President Putin. There is no question of that.

"Whatever the reason for not accepting that, maybe it's like the OJ [Simpson] jury, don't believe in DNA," he added. "Okay, but that's an opinion."

———

Looking ahead, Ledgett believes that, due to its success in 2016, Russia will target future US elections and other critical infrastructure.

"I fully expect that they are. I mean, again, why would their behavior change?" he asked.

Russia has penetrated other critical infrastructure such as power grids, telecommunications systems, and water treatment systems.

Such penetrating attacks give Russia the option of shutting down those systems, in the event of war.

"I think that this is part of a shooting war, if we get to that point, or a really just precursor to a shooting war," said Ledgett. "That countries like Russia, and China, maybe North Korea, maybe Iran, have the capability to get into and affect our critical infrastructure, in things like the electrical power grid, telecommunication system, the financial sector."

However, Russia's preference is to remain below the threshold of a shooting war. Russia's goal is to inflict damage on the United States without sparking retaliation that brings unbearable costs to Russia. The danger for the United States is that the damage so far may be just bearable enough to ignore it. Ledgett uses the analogy of the frog in the pot of boiling water.

Russia is constantly but slowly turning up the temperature, until, at some point, "the frog is boiled," Ledgett said. "So, I worry about us allowing the kind of things that have been going on, to go on. So, we reset the threshold. No meaningful response, nothing that actually deters them. That's the new baseline. That's the new ground floor.

"This was a big win for the Russians," warns Ledgett of the 2016 election interference. "I'm sure lots of folks got medals pinned on their chests, and promotions. If I were their boss, I would've done that, because they did a great job. Low cost, no real impact to Russia, huge payoff, in terms of damaging the status of the United States."

LESSONS

Russia's interference in the 2016 US presidential election proved that a relatively simple influence operation could—potentially—

move an election in the world's most powerful nation, with severe consequences both within the US political system and between the United States and Russia on the most sensitive national security issues. Whether Russia did so with the assistance or knowledge of American citizens, including members of the Trump campaign, remains unresolved. However, President Trump's and some Republicans' equivocal response to Russian meddling demonstrated that the Kremlin would not meet unified American resistance, further emboldening Moscow to attempt similar interference again.

Russia's probing attacks on US political targets in the run-up to the 2018 midterm elections showed that the US response to its meddling in 2016, largely in the form of economic sanctions, may not have imposed costs sufficient to deter Moscow's malign activities against the United States. More recently, President Trump authorized the Pentagon and Cyber Command to respond to foreign cyberattacks with offensive cyber operations of America's own. However, it is not clear what level of foreign interference would trigger such a response. US cyber experts and policy makers agree that maintaining the integrity of current and future elections requires both a credible offensive deterrent and more effective defensive measures. It is not yet clear that the United States has achieved either. Without doing so, there is evidence that not only Russia, but other foreign adversaries, led by China, Iran, and North Korea, have attempted or experimented with similar election interference, undermining the confidence of the American public in the electoral process, a potentially debilitating blow to American democracy that—once ingrained in the public consciousness— would be difficult to reverse.

CHAPTER 8

Submarine Warfare

(RUSSIA AND CHINA)

The landscape of the Arctic Circle is a blue-gray kaleidoscope of broken ice, resembling shards of glass floating on the surface of a dark, undulating sea. This vast frozen expanse is, however, far from frozen in place, constantly readjusted by the rotation of the earth and the melting and refreezing of the ice. The warming climate has accelerated this mesmerizing dance. Each summer, more of the ice cap melts, and each winter, less of it refreezes. Sea ice in February and March 2018 reached its lowest levels ever recorded, about a half million square miles below the 1981–2010 average.[1]

In March 2018, one day before my forty-eighth birthday, I hopped aboard a Twin Otter turboprop in Deadhorse, Alaska—one of the northernmost populated areas of the United States—for a flight to a US Navy camp pitched on the arctic ice not far from the North Pole. My ride that day was one of the older aircraft in the charter fleet, with a wooden floor and a thin metal skin barely insulating passengers and crew from the subzero temperatures outside. A warm day in Deadhorse in March reaches the single digits Fahrenheit. The thermometer at the airport that morning registered just below zero without the windchill.

Our destination was an ice camp that would serve as command-and-control center for the US Navy's 2018 ICEX exercises. Three nuclear attack submarines—two American and one British—would spend the next three weeks under the Arctic training for submarine warfare in the harshest sea environment in the world. A few days earlier, an advance team had carved out a landing strip near the camp, on a spot carefully chosen so that the ice was thick enough to support a three-ton Otter landing at nearly one hundred miles an hour. The safety margin was razor thin: my fellow passengers and I were all weighed holding our gear to make sure the loaded plane did not exceed maximum weight. I tipped the scale at 248 pounds with my parkas, backpack, and sleeping bag.

Ninety minutes out from Deadhorse, the ice camp came into view. Constructed in just a few days, it resembled a mountaineering base camp with a semicircle of rigid tents housing a command post, a mess hall, sleeping quarters, and a "pee tent" in easily identifiable yellow. A Navy dive team stationed there had jokingly placed inflatable palm trees outside their door. Here, the temperature hit minus-forty Fahrenheit, on par with some of the coldest days at the peak of Mount Everest.

From the air, the camp's surroundings appeared stationary, but arctic ice is a jigsaw of interlocking ice floes in constant motion. Our camp was situated atop a four-square-mile ice floe, migrating east-southeast at a half-mile an hour, about twelve miles a day, in waters ten thousand feet deep. The ice here is what's known as "multiyear ice"—born at the North Pole and formed over the course of several years as it continues its long migration to its eventual demise in the warmer waters to the south. Over its life span, the salt from the seawater gradually leaches back into the sea, leaving freshwater ice that is an enchanting crystal blue. Arctic ice is a living organism, always in motion, always being born and dying. Here nothing has ever been permanent, but as the planet warms, the life cycles are getting shorter and presenting new challenges for US submarine forces.

US Navy engineers surveilled this part of the Arctic for weeks searching for ice thick enough to support the ice camp. Weak ice can be deadly. During the 2016 exercises, a crack appeared in the middle of the camp within minutes, forcing an evacuation. Choosing the right location is a Goldilocks problem—thick enough to support the camp, thin enough to allow US submarines to break through the ice when surfacing.

Our Twin Otter made a wide circuit of the camp before settling in on approach. I expected a slippery landing on my first ice runway, but the powdery snow on the surface slowed the plane down quickly. As the crew opened the door, an arctic blast hit me immediately. The bare strip of skin between my ski goggles and face mask—the only exposed part of my body—went painfully numb in seconds.

Outside, the Arctic looked to me like the surface of an alien planet. The sun was blindingly bright, reflecting off the endless expanse of white. A steady wind carried wisps of snow across the ice, whistling so loudly I could barely make out voices. I was wearing multiple layers of military-grade polar gear, but the cold attacked my extremities quickly. I imagined the first explorers trudging through here in leather and fur and marveled at the combination of toughness and ambition that kept them going.

In the distance, our ride for the next few days peaked up

through the surface. The USS *Hartford*, a *Los Angeles*–class nuclear attack submarine, had broken through the ice earlier in the day. Only its monolithic black conning tower, seemingly disembodied from the rest of the boat, was visible. *Los Angeles*–class subs are some of the biggest in the world, but in the arctic landscape, the *Hartford* looked minuscule.

From the frozen surface of the Arctic, the giant metal cigar crammed with a nuclear reactor capable of powering a small city suddenly looked inviting. I climbed down the hatch and into the warmth below. Embarking on a submarine brings you into a world impervious to its surroundings. Like a Las Vegas casino, a sub's interior gives no sense of the time or conditions outside. Even at speed, you sense movement only on tight turns or steep ascents and descents. There is a quiet mechanical hum in the background and an oddly uniform temperature of about 68 degrees whether the water temperature outside the hull is 80 degrees in the Caribbean or 31 degrees in the Arctic. It is also counterintuitively dry. A submarine acts like a natural condenser, leaching the moisture from inside the sub to outside the hull.

It is immediately clear, however, that this is a weapon of war. A modern nuclear attack submarine is an engineering marvel of devastating proportions. The nuclear reactor pumps out 165 megawatts of power for more than thirty years without replacing its fuel. The sub's arsenal includes four forward-facing torpedo tubes and twelve vertical missile launch tubes, giving it the ability to strike targets below and above the surface and on land.

With weapons and propulsion systems taking up more than half of the interior, there is little space left for the crew. For the 150 submariners on board, there are fewer than 120 available bunks. That means crew members with the lowest rank and shortest time of service face the indignity of what submariners call "hot-racking," that is, sharing the cramped, casketlike bunks in eight-hour shifts. As a guest on board, I was assigned a bunk of my own, though at six foot three, my head and toes touched either end and my nose was fewer than six inches from the bunk above me. There's a reason submariners compare these beds to caskets.

Since space is at a premium, none is wasted. Under every seat, below every bunk, behind every wall is a compartment for something. Even the space under the mess hall seats is reserved for condiments, each compartment dutifully marked "ketchup and mustard," "barbeque sauce and A1," and so on. Moving through the corridors and up and down the stairways of the sub's three decks requires an astute sense of personal space. Whenever you stop in place, even in the bathroom, you will soon find yourself in somebody's way. And so you have to be ready to move quickly in tight quarters. The interweaving movements of a submarine's crew is a delicate dance of shimmying to the left and right and backward and forward. Somehow these sailors manage to pull it off with a smile and a polite nod.

Their diligence fits their unique mission. Submarine forces are called "the silent service," both for their ability to sneak up on adversaries and their characteristic humility in the face of brutal conditions and long, isolating deployments at sea. During a six-month tour, a sub might surface for fewer than ten days. That means long stretches without any contact with the outside world: no emails from home, no Skype calls with the kids.

That ability to disappear is part of the job. The submarine fleet's mantra is: "On scene, unseen." Through its combination of range, endurance, and silence, submarines have a unique ability to project military power anywhere in the world, especially in the Arctic. For most of the year, this is still a region accessible only by submarines and icebreakers. The US Navy does not have a single icebreaker in its fleet, unlike the Russian navy, which has dozens. The US Coast Guard does have three icebreakers. However, submarines remain the best option for the US Navy to operate in the Arctic and, if necessary, wage war there.

Submarines have an outsize role in America's nuclear deterrent. Today US Navy submarines carry some 70 percent of accountable US nuclear warheads, unlike Russia, which deploys the bulk of its nuclear weapons in land-based missile silos. While land-based silos can be targeted and destroyed, submarines are constantly in motion and virtually invisible to the enemy. That means, in theory, that US submarines can deploy within missile range of enemy territory without warning and unleash nuclear Armageddon.

The destructive power of a single ballistic missile submarine unsettles the imagination. *Ohio*-class ballistic missile submarines carry twenty-four Trident II submarine-launched ballistic missiles (SLBMs). Once in the atmosphere, traveling at twenty-four times the speed of sound, the Trident divides into eight separate warheads, giving it the ability to strike eight separate targets. Each warhead delivers the destructive power of up to 475 kilotons, or 30 times the power of the atom bomb that destroyed Hiroshima. That means a single SSBN submarine has the power to wipe out some 200 Hiroshima-sized cities.

What is the USS *Hartford* doing here on the top of the world? Why is the US Navy deploying two of its most powerful submarines to the Arctic? And why now? It's immediately clear this is no second-tier mission.

To leave no doubt about the seriousness of the mission and the message, ICEX is a live-fire exercise. By the end of the live-fire portion of the operations, dubbed "TORPEX" for torpedo exercises, these submarines had fired four torpedoes at targets.

Just as with navigating under ice, waging war under ice brings particular and difficult challenges. In the arctic environment, the *Hartford*'s sensors, while very capable, can still mistake an enemy sub for the ice around it. During TORPEX, the US Navy calls on its submarine officers and crews to prove they can see the difference—and take out enemy subs.

"It's just reminding everyone that this is different," said Captain Matthew Fanning. "We're underneath the ice and we need to be cognizant of where we are and that we don't have access to GPS easily. We don't have access to communications easily. So that does change the way we think about our operational pattern."

For the submariners of the USS *Hartford*, torpedo exercises are deeply connected to their history. The first US naval vessel to bear its name was the Civil War–era USS *Hartford* steam- and sail-powered sloop. And it was aboard that *Hartford* as it battled Confederate forces in the Gulf of Mexico that Admiral David Farragut uttered his famous order, "Damn the torpedoes. Full-speed ahead."

In TORPEX, task number one was identifying, targeting, and destroying what submariners call an "ice-pick sub," that is, an enemy submarine deliberately hiding up against the surface of the ice, where a submarine's multiple tracking systems find it hard to distinguish an enemy or friendly vessel from the ice itself. The movement of the ice—the collisions and cracks and submersions—also generate background noise that helps mask the sound of adversary submarines, which are already virtually silent.

"I think it's important that we continue to prove that we are as capable as we have always been," Captain Fanning told me. "Doing these exercises gives us a chance to practice all the things we believe work in a different environment up here—in literally one of the harshest environments on the planet.

"Twenty-eight-degree water—thirty-one degrees right now

looking at the gauge—really does change how we operate the submarine," he said.

A submarine's field of vision relies on four separate sensing systems. There is a top sounder, which sends radio waves upward to build a picture of the underside of the ice. There is side-scan sonar, which paints a picture of the surroundings to port and starboard. A new addition is a live-stream video camera, which provides live video images of the surface.

Those sensors give a great deal of vision but far from perfect situational awareness. And since they are underwater, they cannot track their own position via GPS. Signals from GPS satellites do not penetrate water or ice.

In the exercises, the submarines play a game of hide-and-seek. Two attack submarines chase a target submarine known as the "rabbit."

Playing this game under the Arctic—indeed all submarine operations in this environment—poses unique challenges.

Arctic ice looks level from above. The surface facing up toward the air and sky is, with the exception of cracks and low snowdrifts, largely flat. But underneath, it is like a vast upside-down mountain range. Where the sections of ice meet, they propel huge sheets of ice downward. These create giant ice keels extending down from the surface. They present enormous danger to submarines. A collision with one of them at a sub's cruising speed of twenty knots can disable equipment, injure the crew, and, at worst, cause fatal cracks in the hull.

"The difficult part is that you got a roof over your head now," said Rear Admiral James Pitts. "The ice canopy. You have potential of ice keels that hang down in the water—they can reach down there about a hundred, hundred and fifty feet."

Ice keels also confuse a submarine's weapons systems. Torpedoes can easily mistake one for the hull of a ship or sub.

On the morning of March 9, 2018, the crew of the USS *Hartford* spent hours tracking the "rabbit" submarine under the ice. Despite all the advanced sensory systems, it is an imprecise art. US submarines run quietly, and the ice keels and other features provide additional cover. Looking at the monitors myself, I could not spot a single identifiable blip that resembled the target sub.

When they were confident they had found their "rabbit," the Fire Control Team prepared targeting options for the commanding officer. There are no straight shots underwater when both the targeting sub and its target are moving. Weapons officers also have to factor in currents and, as for the sub itself, differences in water salinity that will affect the torpedo's buoyancy.

In preparation, down in the torpedo room, weapons officers flooded torpedo tubes one and two. The tubes must be flooded with water from an internal tank as a first step to prepare for firing. Prior to flooding, seawater pressure, bearing down on the hull with the force of forty-two pounds per square inch, prevents the outer tube door, or muzzle door, from opening. Once flooded, the pressure equalizes, allowing the door to open and putting the torpedo in position to fire.

Within minutes, the weapons officer shouted, "Stand by, Tube One!"

And then, "Fire Tube One!"

An audible hiss burst from the torpedo tube as a rush of compressed air propelled the torpedo forward. As it did so, I could feel a change of pressure throughout the submarine, enough to pop my eardrums.

Like everything on a submarine, the firing mechanism is precise. Just enough air is expelled to send the torpedo toward its target, though not enough to fill the tube entirely. The rest of the air is captured and vented off inside the sub as the muzzle door closes behind the torpedo. Otherwise a bubble might rise to the surface, giving away the submarine's position. The tube is then quickly filled with seawater to compensate for the lost weight of the torpedo, so as not to disturb the level and quiet sailing of the sub.

Now came the wait. Their target, the "ice-pick sub," was some two miles away. The exact speeds of US Navy torpedoes are classified. However, commanders say they are "in the neighborhood" of twice the speed of the submarine itself. Those speeds are classified as well but with sub speeds widely estimated at greater than 20 knots, torpedoes likely travel in excess of 40 knots, close to 50 miles an hour.

As we waited, Captain Fanning told me. "This is our one opportunity to see if what we've figured out works out in the middle of the ocean still works up underneath the ice."

Seconds later, the Fire Control Team announced they had hit their target. The USS *Hartford* had "killed" an enemy submarine. And it was now acquiring a new enemy target: a second rabbit a mile off its port side. Again, the weapons officers developed a fix on the enemy submarine.

"Stand by, Tube Four!" shouted the weapons officer.

And finally, "Fire Tube Four!" A second enemy sub was dead.

The shrinking of the ice has transformed the Arctic from a wasteland to a land of opportunity and potential conflict. Vast oil resources that were untouchable are now within reach. A northern sea route, the dream of seafaring nations dating back to the eighteenth century, is becoming a reality. And what is newly open to commercial traffic is also open to warships and submarines.

With US and Russian territory separated by just a few dozen miles at the closest point, this front of the Shadow War puts the two adversaries in close-quarters combat. This is a new Great Game.

The ICEX exercises are intended to leave no doubt about America's arctic ambitions. These war games had tailed off after the end of the Cold War but in recent years have been ramped up again. Great Britain, which had stopped participating in 2010, rejoined in 2018. The United States is not the only Western ally recalibrating its national defense strategy for the Shadow War.

Rear Admiral James Pitts, then commander of the US Navy's Undersea Warfighting Development Center, met me at the camp. The UWDC trains the Navy for submarine warfare. And as Russia expands its submarine forces in the Arctic and around the world, the ICEX exercises are more important than ever.

"When you look at our national defense strategy," Pitts told me, "we are well aware that we are in a great-power competition environment. The Arctic is one piece of that—all the more reason why we the Navy are practicing up here and holding exercises to make sure we can operate effectively."

Today US submarine commanders are confident the United States retains the advantage over Russia below the surface. However, they acknowledge that advantage is shrinking.

"We have an advantage, but our competitors are trying to catch up faster and faster," said Pitts. "So, our role at the Undersea Warfighting Development Center is to keep our advantage as much as possible—and getting our undersea war-fighting forces ready to maintain that superiority.

"Our higher-level guidance—from the secretary of defense, secretary of the Navy, and the CNO [chief of naval operations]— clearly spells that out for us," he added. "We are aligned and moving out on what the Navy believes is the Navy the nation needs."

As to whether Russia was watching his submarines during ICEX, Pitts answered with a submariner's understated bravado: "If they are, you know, so be it. It doesn't bother us."

Submarines are the tip of the spear of this new Great Game. *Los Angeles*–class attack submarines like the USS *Hartford* were designed at the peak of the Cold War with the Soviet navy in mind. Their mission: to hunt and kill Soviet submarines and surface warships. As the Soviet Union collapsed, however, and the Russian threat to the United States subsided, they became, in the view of some submariners, an exquisitely advanced warship without a clear mission. In the intervening years, they dabbled in counterterror missions, firing cruise missiles at targets in Afghanistan and Iraq. They were also modified to deploy Navy SEAL teams. Today, however, they are training once again for their original mission of tracking—and, in the event of war, destroying—Russian subs and warships. There is no shortage of submariners who welcome the return to their roots.

The ICEX exercises are intended to demonstrate that US Navy submarines like the *Hartford* can execute this mission in the Arctic. And the intended audience is Russia, whose northern coastline was just a short flight from the ice camp.

"We border on this ocean. And it's of strategic interest to us," Commodore Ollie Lewis, commander of the USS *Hartford*'s squadron, Submarine Squadron 12, based in New London, Connecticut, told me. "We see it as a global common. We see it as a passage that we'll rely upon at any time in order to flow submarine forces. But it's critical to our strategic advantage to be able to use it—and be proficient in using it—whenever we need to.

"The US is an arctic nation," he said.

Russia, however, is an arctic nation as well, though with 50 percent of its coastline on the Arctic Ocean, it has far more presence and dependency. For Russia, the Arctic isn't just strategically valuable. It is a matter of survival. And so Russia views the region as sovereign territory.

To make that position clear—and to protect its northern flank and project power from it—Russia has formed an "Arc of Steel" spanning its entire northern coastline. It consists of more than four dozen airfields, ports, and missile defense systems, supplemented by deployed troops, surface ships, and, of course, submarines.

Russia has made other more symbolic shows of force. In 2007, Russia deployed two submarines two and a half miles below the North Pole to plant a meter-high Russian flag on the seabed. It was a record-setting dive—and one with a message. The flag was planted on the underwater Lomonosov Ridge, which the Kremlin claims is linked to its continental shelf, thus, under international law, giving Russia rights to an additional half a million square miles of arctic territory and the rich oil reserves underneath.[2]

In 2015, Russia made an even more aggressive demonstration of its arctic ambitions. On April 20, Russian deputy prime minister Dmitry Rogozin landed on the archipelago of Svalbard and posted a picture of him and his team there on Twitter with the pronouncement: "The Arctic is Russian Mecca." Trouble is, Svalbard is part of Norway—as much as Alaska is part of the United States.

Norway summoned Russia's ambassador to register its alarm but neither Rogozin nor the Kremlin were deterred. In fact, Rogozin has been one of the most vocal supporters of Russia's arctic vision, framing his country's claim in historical terms. The year before his Svalbard adventure, Rogozin wrote that in pursuing its arctic claims, Moscow was merely righting the mistakes of the past. He even went so far as to claim that Russia retains a historical right to Alaska.

"Russia giving up its colonial possessions makes it necessary to look in a different way at our diplomacy in the era of Gorbachev and Yeltsin, trading away pieces of the Soviet Empire," he wrote.[3]

This is a consistent message and a driving force of Vladimir Putin's approach to the Shadow War. He sees Russia as righting historical wrongs, restoring his country to its rightful position as the dominant power in its "near abroad" and to more equal footing with the United States globally. For Putin, the principal obstacle to all those ambitions is the United States.

Russia's show of force under the surface extends far beyond the Arctic. Today the Russian navy is deploying its nuclear attack and ballistic missile submarines in numbers, range, and aggression not seen since the Cold War.

"NATO is viewed as an existential threat to Russia, and in the post–Cold War period, the expansion of NATO eastward closer to Russia and our military capability they view as a very visceral threat to Russia," now-retired admiral Mark Ferguson, then commander of US naval forces in Europe, told me in an interview in April 2016.

"They have increased the readiness levels of the [submarine] force," Ferguson said. "They are operating it with more . . . out-of-area deployments, and what we are seeing is an improvement in the readiness of that force as well."

Moscow has built and deployed entirely new categories of submarines with far advanced capabilities. Its new subs are quieter and better armed and can operate with greater range.

"The submarines that we're seeing are much more stealthy," Ferguson said. "We're seeing [the Russians] have more advanced weapons systems, missile systems that can attack land at long ranges, and we also see their operating proficiency is getting better as they range farther from home waters."[4]

The vanguard of this new Russian navy force are the Project

955 *Borei*-class nuclear ballistic missile submarines and the Project 885 *Yasen*-class nuclear attack submarines. It is their quietness that most concerns US Navy commanders. Stealth is submarines' chief advantage. A quiet submarine can turn up off an enemy's coastline and rain down nuclear warheads with just a few minutes' warning.

To make that point clearer, Russian submarines are regularly turning up in places they have not since Soviet times. Russian submarines are swarming down through what US Navy commanders refer to as the Greenland-Iceland-UK gap, or "GIUK" gap, the section of the North Atlantic that leads from Russian bases on its north coast down into the Atlantic, providing access to Western Europe, America's East Coast, and beyond.

Russia has newly stationed six submarines in the Black Sea, giving it increased access to the Mediterranean, crucially, from a warm-water port that is operational year-round. Russia is also opening a new submarine base just up the coast from Crimea in the Russian port of Novorossiysk. When the base construction was announced in 2016, it was not lost on Russia experts that "Novorossiysk" means "New Russia."

Since its intervention in Syria in 2015, Russia has reopened its Soviet-era naval base in Tartus, once the home of the Soviet Union's Fifth Mediterranean Squadron. At its reopening, the Russian state television network RT reported, "Russia is allowed to bring in and out any kind of 'weaponry, ammunition, devices, and materials' to provide security for the facility staff, crew, and their families throughout the territory of the Syrian Arab Republic 'without any duties or levies.' "[5]

Among the vessels Russia has deployed to Syria are *Kilo*-class attack submarines.

In addition, the United States has observed a significant increase in Russian submarine activity in the eastern Mediterranean, including nuclear attack submarines from its North Sea fleet, as well as the launch by Russian submarines of Kalibr missiles into Syria from the Black Sea.

US Navy commanders believe Russia's naval expansion is intended to deny NATO alliance members, including the United States, the ability to operate in Russia's "near abroad."

"What we see in these operations is that they are focused on protecting the maritime flank of Russia but also denying NATO the ability to operate within the maritime flanks as well," said Admiral Ferguson.

"And I'm speaking of the Baltic Sea, Black Sea, and areas in the far North Atlantic around Norway," he added.

Each of those areas borders the territory of NATO alliance members, including the Baltic nations of Estonia, Latvia, and Lithuania, as well as Norway and Great Britain. Russia's new submarine deployments are a direct challenge to NATO countries to which the United States has treaty obligations to defend in the event of war.

Russia's show of force extends to the air as well. US air and naval forces have encountered increasingly aggressive activity by Russian military aircraft. A Russian jet's flyby of the USS *Donald Cook* in international waters in the Black Sea in 2016— approaching within thirty feet laterally, and one hundred feet vertically—was the closest such interaction US forces had seen since the Cold War.

"While we have seen these interactions before, this one was different due to the proximity to the ship, the altitude, and the flight path," said Admiral Ferguson. "We had radio calls in both English and Russian and the aircraft didn't respond and proceeded on a course directly at the ship."

Most alarmingly, Russian submarines are turning up more

and more off the US coastline. The apparent message: Russia's advanced submarines can strike the US homeland with minimal or no warning.

Russia's expanded activities in the Arctic and around the world—and its more advanced, quieter submarines—have rekindled a sense of urgency across NATO. For the United States and NATO, the renewed focus on its old Cold War adversary, Russia, and old Cold War weapons and missions represents a startling reversal. After the fall of the Berlin Wall in 1989 and collapse of the Soviet Union two years later, NATO drastically reduced its military capabilities. In doing so, European nations were mirroring Russia's own reduction in forces and activities. This has been a consistent error of the Shadow War by the West: imagining that Russia shared the West's ambitions when Moscow's behavior showed it did not.

"After the Cold War, NATO reduced its naval capacities and especially antisubmarine warfare capabilities. We also did less training, so both the capabilities and the skills were reduced," NATO secretary general Jens Stoltenberg told reporters in a background briefing in December 2017.

Beyond shrinking its military forces, NATO radically changed its mission. Rather than defending Europe, the alliance redirected its ambitions and forces on projecting power beyond its borders. The most dramatic demonstration of this new focus came after the September 11, 2001, attacks on New York and Washington, when NATO allies for the first time invoked Article 5 of the NATO treaty, which calls on allies to come to the defense of a member nation under attack by an adversary. NATO forces deployed to Afghanistan in November 2001 to fight this new enemy—Al

Qaeda—carrying the alliance several hundred miles from its easternmost border.

"From 1949 to 1989, NATO did one thing and that was collective defense in Europe, deterring the Soviet Union," Stoltenberg said. "Then the Berlin Wall came down, the Cold War ended, and NATO reduced its focus on collective defense in Europe. We transformed into an alliance which was focused on projecting stability beyond our borders, fighting terrorism in the Balkans, Afghanistan and so on."

The clues to Russia's military expansion came long before it increased its submarine activity, even before the 2014 annexation of Crimea and invasion of Eastern Ukraine. In fact, the aftermath to Russia's military intervention in Georgia in 2008 gave the first clear signs of Russia's new military footing.

What followed was Russia's first significant military expansion since the end of the Cold War.

"After Georgia, and since 2008, there has been a significant modernization of the Russian capabilities, including naval capabilities," said Secretary General Stoltenberg.

Again, as with Russia's invasion of Ukraine and increased cyberattacks on the United States and Western Europe, 2014 proved to be a crucial turning point. NATO and US commanders have seen a particular focus on submarine capabilities. Russia has deployed thirteen more submarines since 2014—a dramatic expansion in such a short time frame. Since its annexation of Crimea, Russia has transferred six advanced, modern missile submarines to its Black Sea Fleet. And the Russian navy has tested and deployed new weapons on those new submarines, including its Kalibr missile.

The Kremlin has deployed those new, more advanced submarines in areas in which Russia had not been active since Soviet times, including operating again in the Mediterranean and off the East Coast of the United States.

"They have activities which they ceased after the Cold War which have now started again," said Stoltenberg. "So, this is partly about more [submarines] but also about new kinds of activities or activities that haven't taken place for many years."

Like the United States, NATO has responded with a show of force of its own. However, NATO leaders are increasingly speaking about Russia in existential terms. For NATO, Russia's expansion is a threat to Europe's survival.

"NATO is now in the process of implementing the biggest reinforcement of collective defense since the end of the Cord War," said the NATO secretary general.

"There is a new challenge out there, NATO is responding," he said.

NATO began the modernization at the NATO summit in Wales in 2014. Russia's invasion of Ukraine—after initial foot-dragging—finally jolted the alliance into action. And alliance military planners took an immediate focus on submarines.

"We identified antisubmarine capabilities as one of the gaps we had to fill," Stoltenberg told reporters, "and we have gradually then strengthened our submarine and anti-submarine warfare capabilities with more submarines, planes, ships which can detect and track submarines with more skill, meaning more exercises."

The United States for the first time deployed to Europe its P-8 Poseidon aircraft, the world's most advanced aircraft for both tracking and—in the event of war—destroying enemy submarines. NATO ally Norway, which lies close to one of Russia's most crucial submarine bases, contracted to buy P-8s of its own. NATO is also conducting more exercises to hone skills in antisubmarine warfare. "Dynamic Mongoose," a submarine warfare exercise in the North Atlantic, is now an annual event. NATO conducts a similar annual exercise, dubbed "Dynamic Manta," in the Mediterranean.[6]

Most important, NATO refocused its energy and resources on its old Cold War mission of defending the very continent from Russia. Stoltenberg identified 2014 in particular as a "pivotal year in the history of NATO and our security.

"We had to come back home and once again focus on collective defense in Europe," he said. "The challenge is that we cannot stop managing crises beyond our borders, so for the first time in our history we need to do crisis management beyond our borders, and at the same time focus on stepping up our efforts for collective defense in Europe."

NATO leaders caution they have no interest in going to war with Russia. In fact, they argue that reinvigorating a strong collective defense is the best way both to deter Russia and to diminish the chances of miscalculation and conflict.

"The best way to prevent a conflict is to send a clear signal to any potential adversary that we are able to defend and protect any ally, and therefore we need to have a strong force to be able to send that message," Stoltenberg warned.

NATO leaders speak of allies' Article 5 commitment as the bedrock of their collective defense. For Europeans, who live within range of Russian tank columns, Article 5 is sacrosanct.

"NATO allies are covered by Article 5," he added. "They are covered by an ironclad strong commitment from all allies to defend each other."

But that commitment faces a new challenge as well. President Trump is the first American leader since the alliance's founding to publicly question America's commitment to defend NATO allies if they are attacked. Trump's omission, while dismissed by his defenders as a negotiating tactic in the perennial NATO budget battles, reverberates through NATO capitals to this day.

Now lurking under the waves is another challenger for US submarine dominance: China. In the last two decades, China has made enormous strides, growing its naval forces above and below the surface at a rate matched only by the rapid US naval buildup in World War II.

"Their trajectory is very fast. It's a high ramp," said Commodore Ollie Lewis, commander of the *Hartford*'s submarine squadron. "Their build rate is incredibly high; their proficiency is rising" (see pages 133, 135–36).

To get perspective on the speed of China's rise, start with the numbers. According to data from the International Institute for Strategic Studies (IISS), in 2000, China's People's Liberation Army Navy (PLAN) deployed a total of 163 surface ships and submarines versus 226 in the US Navy. By 2016, China had narrowed that gap to near parity: 183 ships and submarines in the PLAN compared to 188 in the US Navy. By 2030, US military planners project that China will surpass the US Navy, at least in raw numbers, with 260 ships and subs versus 199 for the United States.

"They clearly recognize the value of a modern navy and the influence it brings," said Lewis. "They're going to be a competitor for a long time to come and we're ready to compete alongside them in order to maintain the balance we're looking for."

ICEX is part of the US Navy's response to its new challengers. And for many submarine officers, the return to a Cold War footing is a welcome change. The USS *Hartford*'s commanding officer, Captain Matthew Fanning, graduated from the US Naval Academy in 1999 and received his first submarine deployment on the USS *Los Angeles*—the namesake of the *Hartford*'s class—in 2001, eight months before the September 11 attacks.

"I started my career in the counterterrorism mind-set in the wake of 9/11 and what that meant for the submarine force," Fanning recalled.

For him, counterterrorism was an unnatural mission for submariners.

"Now, it is back to the lethality. Our primary mission of the submarine force is to be able to leverage our offensive weaponry, like a torpedo, against a threat. So, there's been a shift in emphasis on to our ability to do that," said Fanning.

Speaking to him in his wardroom, I could sense both his excitement and sense of comfort. He is and always has been a submariner, and now he is doing what he believes submariners—and the submarines they serve on—were meant to do.

"It's more what we believed when we designed this particular ship," said Fanning. "I think we believed the submarine was capable of all kinds of mission sets. But this feels definitely more natural. I spent most of my time earlier in my career in very, very shallow water conducting operations in preps to conduct those types of strikes.

"Out here in deeper water," he added, "it definitely is what we were made to do."

———

Russia is America's closest competitor below the surface, making progress not only with more advanced submarines but entirely new submarine platforms. In March 2018, Russian president Vladimir Putin boasted of a drone submarine capable of carrying a nuclear weapon across oceans to attack enemy cities. In a brash speech weeks before his election to a fourth term, Putin held court before the Russian parliament in front of giant floor-to-ceiling video screens displaying computer animations of the new weapons systems. In one, an unmanned underwater drone, launched from a submarine, carried a nuclear warhead underwater at high speed before emerging to strike a coastal city.

"Russia has remained a major nuclear power," Putin said. "No, nobody really wanted to talk to us about the core of the problem, and nobody wanted to listen to us. So listen now."

The Russian president also revealed what he claimed was a nuclear-powered cruise missile with an unlimited range and the ability to weave around enemy missile defenses, and a new missile capable of hypersonic flight, traveling—he said—"like a meteorite, like a ball of fire" at several times the speed of sound. Menacingly, Putin claimed the new weapons would render both US and NATO defenses "completely useless."[7]

Russia's advances are changing the way submarine commanders and their superiors are thinking about and planning for a potential conflict with Russia.

"They're definitely upping their game and we're upping ours, too," Commodore Lewis told me. "We're never gonna get better at this standing still. The competition has started and we're not going to sit back and rest on our prior advantage."

As the United States adapts and expands, Russia is using submarines for entirely new hybrid warfare missions: blending submarine warfare with cyber warfare in the event of all-out war. The Russian navy has modified ballistic missile submarines to be able to carry and deploy smaller, deep-diving subs. Capable of descending as far as several thousand feet, the subs can reach the ocean floor and, once there, perform a range of tasks. For several years now, as Russia has deployed and tested these new submarines, the United States has been watching.

"We are now seeing Russian underwater activity in the vicinity of undersea cables that I don't believe we have ever seen," US Navy rear admiral Andrew Lennon, the commander of NATO's submarine forces, told the *Washington Post* in December 2017. "Russia is clearly taking an interest in NATO and NATO nations' undersea infrastructure."[8]

These Russian submarines have conducted multiple missions in the Atlantic locating, monitoring, and in some cases manipulating and moving the undersea cables that carry the bulk of communications between the continental United States and Europe.

Another Russian submarine—the *Yantar*—has been modified to carry two smaller submarines capable of manipulating or potentially cutting undersea cables. The *Yantar* is based at a Russian port near Norway and has been observed conducting similar activity at cables laid on the floor of the North Atlantic.[9]

A third enlarged and modified Russia submarine was deployed in 2017 for activities in the Arctic. Russia's *Izvestia* magazine reported at the time of the submarine's launch that it "will study the bottom of the Russian Arctic shelf, look for minerals, as well as placing out submarine communication systems."[10]

US military officials believe that by "submarine communication systems," the Russian navy really means this sub will deploy a new submarine detection system on the floor of the Arctic, designed to locate and track US submarines.

A professor at Russia's Academy of Military Science, Vadim Kozyulin, nearly acknowledged as much, telling *Izvestia*, "The submarine will provide a global deployment of underwater monitoring system, which the military is building at the seafloor of the Arctic Oceans."[11]

He called the *Belgorod* "the most unique submarine in the Russian Navy."

The United States is not certain of Russian intentions, but intelligence and military officials suspect that Russia is perfecting the ability to cut or otherwise disable and restrict these lines in the event of conflict with the United States. This would have an immediate and devastating impact on US civilian, military, and government communications. The international financial system depends on these lines being open.

As they assess this new submarine threat, US and NATO military commanders speak in terms of war—and of grave threats to Europe and the United States. The alliance is meaningless if Europe and the United States are cut off from each other. Their strength depends on keeping both the lines and lanes of communication open between the North American and European continents.

"We are a transatlantic alliance and to move forces, equipment across the North Atlantic, it is key for a North Atlantic alliance, for NATO to connect North America and Europe. But also of course to be able to have safe and open lines of communication," Secretary General Stoltenberg said.

Russia's newest, most mysterious submarine is an explicit threat to those lines and lanes of communication—and therefore a direct threat to the alliance and its member nations as well.

———

Submarines, in the view of many US commanders, are America's best chance to fight back against all these challenges.

"The role of the submarine forces in the opening stages has become more critical," said Commodore Lewis. "Submarines are going to be able to get to places and conduct action that other units will not be able to, right off the bat.

"We're now going to need the submarine forces *to kick the door* and other forces flow through behind," he said.

As a result of this growing dependence, submarines are a uniquely prized possession for the heads of US military commands around the world.

"Combatant commanders judge their importance based on the number of subs," said Captain Fanning. "We are now back to the blue-water operations we were trained for as young officers."

US submarine forces are deploying new weapons of their own for this new age of submarine warfare. New drone submarines—UUVs, or unmanned underwater vehicles—are vastly expanding the reach and capabilities of US submarine forces. The USS *Hartford* is able to carry UUVs on board, including versions that can be launched through both its forward-facing torpedo and upward-facing missile launch tubes.

"It's exploding," said Captain Fanning.

And like their larger forebears, UUVs can perform multiple functions, from tracking enemy subs to patrolling coastlines and carrier groups and delivering weapons. The Navy envisions a not-too-distant future in which manned submarines provide command and control for swarms of drone subs.

The United States currently has 53 submarines in its inventory, but because of decommissioning and budget decisions, the US Navy says that figure will drop to 41 by the late 2020s. The 2019 military budget calls for the addition over time of 15 new nuclear submarines, as part of the Pentagon's plan for a 355-ship Navy. The timeline for the expansion currently extends to the middle of the century; however, the Navy says that can be accelerated.

"We cannot maintain 100 percent awareness of Russian sub activity today," retired admiral James Stavridis, a former NATO supreme allied commander and current dean of the Fletcher School at Tufts University, told CNN. "Our attack subs are better, but not by much. Russian subs pose an existential threat to U.S. carrier groups."[12]

Training in the Arctic is one front in the US Navy's effort to remain ahead.

"We're shifting to an era of great-power competition—and being able to operate here and know the environment, our ability to adapt to it, that's what's adding a new level of urgency to it," said Lewis. "In every case, they're trying to get faster and better at what they do. It's really set them on a ramp where if we don't do the same, we'll find ourselves in a place of falling behind."

O f all a US submarine's operations under the Arctic, surfacing through the ice may be its most delicate and precise. At 18:13 Zulu time on March 10, the USS *Hartford*'s captain and crew prepared to perform their second surfacing of these ICEX exercises. This happened to be my birthday, and because less than an hour before the cook in the officers' wardroom had presented me with a homemade chocolate birthday cake, I thought to myself that this next maneuver made for a pretty good birthday present.

Surfacing from under the ice is very different from the dramatic images you may have seen of submarines bursting through the waves like a cork popping into the air. Those so-called emergency blows are reserved for crises underwater and, of course, are only possible in warm waters. At high speeds, arctic ice can be as impenetrable as cement, so surfacing in the Arctic requires a slower, more choreographed ascent.

As foreboding and bleak as the surface of the arctic ice is, it is the area below the ice that poses the greatest challenges for US submarine forces. The bottom of the ice is an upside-down forest of those feared ice keels, some extending as deep as two hundred feet. Sailing and fighting under the arctic ice is like maneuvering through a massive ice cavern, with stalagmites lurking at every turn.

Arctic waters create other unique challenges as well. Older sea ice slowly leaches its salt away. Multiyear sea ice contains some of the purest water in the world, a quality that is visible in the perfect aquamarine tone of exposed ice sections. This freshwater ice then melts at the surface, creating a layer of water with little to no salinity at the shallowest depths. Fresh water is less buoyant than salt water and as the sub moves between the layers, it must constantly adjust its own buoyancy to maintain steady sailing.

The greatest challenge of the Arctic for submarine crews and commanders is how they respond to emergencies. In the open sea, the safest option during an emergency of any kind, from fires to a hull breach, is to immediately come to the surface, where there is easier access to rescue and, in the simplest terms, air. Surfacing is not as easy when several feet of ice stand in the way. With that difficulty in mind, the USS *Hartford* carries a host of alternative sources of oxygen for its crew. Typically, US submarines carry a six-day supply of these canisters. On arctic operations, they carry enough for thirty days. In a severe emergency, however, these crews need to know how to get to the surface as quickly as possible without endangering the vessel or the crew.

The first step is to identify a section of ice with the ideal thickness: thin enough to break through, thick enough to support the weight of the crew and perhaps a helicopter to ferry passengers in and out. For an attack submarine, the Goldilocks thickness is about three feet. With the help of teams on the surface, the USS *Hartford* selected an area with an estimated thickness of two to three feet, dubbing it "Marvin Gardens"—one of several Monopoly-based designations that would be used on these exercises.

At 18:19 Zulu time, the *Hartford* began a series of passes under Marvin Gardens to inspect the location, sailing at a depth of 176 feet, and a plodding speed of just over five knots. There is little margin for error at this depth. Maintaining a steady pace and depth requires a constant adjustment of propulsion and buoyancy. Captain Fanning shouts constant adjustments to the petty officers who operate the sub's helm—one for the sub's rudders and another for its bow planes. *Los Angeles*–class subs are not fly-by-wire. The helmsmen have a physical link to the sub's steering mechanisms.

At 18:28 Zulu time, the *Hartford* passes directly under Marvin Gardens. Its sensors calculate the ice is seventeen inches thick, slightly thinner than the original readings but still within the safe range. Captain Fanning orders another slow 180-degree turn and another pass under the target. At 18:42 Zulu time, as the sub makes a second pass, navigators mark the surfacing location with an X on their screens. At 18:46, the crew verifies distance to target with a single sonar ping. Having watched *The Hunt for Red October* multiple times (one of a handful of Hollywood submarine films that submariners tend to approve of), I expected to hear an audible pinglike "boing." In reality, it sounded more like the crack of a whip.

It was 19:02 Zulu time when Captain Fanning announced over the ship-wide intercom, "Prepare to vertical surface!" We were still sailing steady at a depth of 179 feet, with a slighter faster speed of 7 knots. I scanned the sub's upward-facing camera for a look at the ice we were about to shatter. Underwater, there are far more cracks and separations visible than from overhead. I would feel much less safe stepping out of the sub and onto the ice than I did on my way in. I decided not to think of how well this ice would support our helicopter.

At 19:12 Zulu time, Captain Fanning announced, "The ship is on final approach. All hands are reminded to hold fast when ordered." The captain ordered "All stop" at 19:16 Zulu. The *Hartford* would make a gentle glide in to the surfacing point. Five minutes later, the *Hartford* was one thousand yards out, on final approach. Another reminder over the intercom: "All hands stand fast."

Over the next three minutes, the captain deftly reduced the sub's speed, from 4.5 knots at 1,000 yards, to 3.9 knots at 900 yards, 3.5 knots at 800 yards, 2.9 knots at 700 yards, 2.8 knots at 500 yards. He then shouted, "All stop!" shutting down the sub's propulsion with fewer than 400 yards to go. We still needed to be slower. At 19:31, he ordered the engines reversed at two-thirds power, shouting "All back, two-thirds." We slowed to 1.7 knots. At 19:32, "All back, two-thirds" again. We slowed to 1.1 knots.

The *Hartford* was now ready for the final step. "Ready to vertical surface!" Captain Fanning shouted, repeating three more times, "Vertical surface! Vertical surface! Vertical surface!" Now his teams began a series of "blows," releasing water from the sub's ballast tanks and replacing it with air—steps to "popping the cork" through the ice. "Blow five thousand [pounds]!" "Blow three thousand." "Blow five hundred."

The captain repeated, "All hands, stand fast!"

Now the helmsmen were gradually angling the *Hartford* up at an angle of five degrees. The angle is subtle but standing on the bridge I could feel the floor shift beneath me. My body, somehow pushed by the forces of gravity, angled forward at a slant.

"Blow four thousand!" "Blow ten thousand!" Our depth was now 140 feet, then 130 feet.

At 19:38 Zulu, we were at 120 feet. Through the upward-facing cameras, I could now see bubbles leaving the sub and bouncing off the underside of the ice. "Blow eight thousand!"

The captain issued a final call to the crew to "Hold fast!" and then a final, swift series of blows: 1,000, 10,000, 5,000. At 80 feet and an upward angle of five degrees, two final blows of 5,000 pounds and 10,000 pounds, and then I heard it: a soft scratching noise of the hull against the ice, a rush of air, and a hard push through. The upward-facing camera now showed the blue arctic sky.

Captain Fanning congratulated the crew on the intercom. Surfacing safely through the ice is one of a submarine crew's most challenging operations. Surfacing of course also makes the submarine visible to anyone monitoring the Arctic—and that, for a vessel designed to operate undetected, is very much intentional.

"I think the operations in the Arctic, the exploration of natural resources is a commitment to saying this is our exclusive economic zone and that our submarine force is capable of operating here just as we operate along our east coast and throughout the world," said Fanning.

Today, as US adversaries Russia and China expand the capabilities and operating area of their own submarine forces, sending that message carries growing danger.

"Anytime I leave port in Groton [Connecticut], I assume that I am operating in a hostile environment and it's not too hard to imagine with submarines surrounded by water, it's always a hostile environment for us," he said. "The assumption should always be that there's someone else out there and we need to be able to respond."

LESSONS

Most Americans alive today have grown up in a world in which American military dominance is unchallenged and unchallengeable. That period is over. Both China and Russia have steadily and rapidly expanded their military capabilities with the intention of neutralizing the US military advantage as well as effectively barring the United States from projecting military power inside their respective spheres of influence: for Russia, its so-called near abroad comprising the former Soviet republics; for China, the seas within what Beijing refers to as the "First Island Chain," that is, the area to the west and north of Japan, Taiwan, the northern Philippines, and Borneo. China's and Russia's investments in their submarine forces are a stark manifestation of this strategy: their express mission is to destroy US carrier groups close to their shores, while also increasingly projecting power—and a nuclear strike capability—farther afield and dangerously close to the US homeland.

US military commanders, including those in charge of its still-formidable submarine forces, express confidence that they are up to the Chinese and Russian challenges. However, even they acknowledge that the US advantage is shrinking and will disappear without strategic and technological change on the part of the United States. This change requires faster and quieter US submarines and the capability to track our adversaries' faster and quieter subs. Meeting the challenge also requires new investments and advancements in next-generation weapons systems, such as hypersonic weapons, one area where China may already be in the lead. Investing in old warships, perhaps including even the revered aircraft carrier, is not sufficient to maintain the US advantage. The danger for the United States is the jarring prospect of losing the "new Great Game" now playing out between the United States, Russia, and China from the Atlantic, Pacific, and Mediterranean right to the very top of the world.

CHAPTER 9

Winning the Shadow War

ARE RUSSIA AND CHINA WINNING?

A re Russia and China winning the Shadow War? On the battlefields described in this book, they have occupied new territory, inflicted damage on US and allied forces, and acquired spoils
of war, often in the form of stolen national security secrets. Russia
still controls Crimea and large portions of Eastern Ukraine. China
still controls its man-made islands in the South China Sea and
is expanding its military presence there. Russia and China have
successfully deployed and tested antisatellite weapons that threaten
US assets in space. China's theft of US state secrets and private
sector intellectual property has not decreased in effectiveness. Both
Russia and China have demonstrated the capability to penetrate
US political parties and election systems for the purposes of interfering in the US political process, with worrisome implications
for future elections. In addition, Russia has established a further
military beachhead in Transnistria. And it continues to disrupt

elsewhere in Eastern Europe via nonmilitary means, such as its attempted coup in Montenegro in 2016.

Losses and setbacks in these battles do not mean the United States has lost the larger, global competition with its two biggest adversaries. That competition is ongoing and intensifying. However, failing to turn the tide on these early engagements of the Shadow War has damaged essential US national security interests, weakened the US position in that global competition, and weakened the US position in the event of an all-out war.

US national security officials agree that the United States must find better ways to fight and defend against the Shadow War, to impose costs sufficient to compel Russia and China to change their behavior, and, if possible, to impose costs sufficient to reverse the gains they have already achieved, or to make those gains untenable. The consensus of the current and former national security and intelligence officials I've spoken with is that none of these steps has so far been taken to a degree sufficient to make America safe.

Some make comparisons to the 1930s and America's slow recognition of the threat posed by Hitler as a disturbing precedent for the lack of urgency in fighting America's new and ambitious adversaries today. During World War II, it took Japan's attack on Pearl Harbor to change minds and spark decisive action by the United States. The unique challenge of the Shadow War is that it is designed exactly to avoid sparking a decisive response, by remaining just below the threshold of war—to defeat the United States without a modern Pearl Harbor. Even a bold attack on America's most precious institution—a presidential election—fell below that threshold and—to the delight of the Kremlin—succeeded in further dividing the United States, rather than uniting US leaders and the US public to take action.

I asked several leaders with direct roles in formulating US and

Western national security strategy for their recommendations on how to win the Shadow War. Together, former director of national intelligence James Clapper, former secretary of defense Ashton Carter, former director of the CIA and NSA Michael Hayden, and former MI6 chief John Scarlett, have more than 150 years of collective experience defending the West against threats at home and abroad. All are "big thinkers" in the US and European national security communities, open to finding solutions both inside and outside established thinking. In fact, they agree that winning the Shadow War requires responses and solutions that fall outside the normal playbook.

On the first, most basic question, they are largely in agreement: the United States is losing the Shadow War, based on the rules of the game as established by Russia and China.

"We kind of are, I think," General Clapper told me bluntly. "When they find ways to push the envelope at levels which they understand are below that which would induce a kinetic, or even mostly kinetic response, if they assess we are not going to push back.

"So when the Chinese do things in the South China Sea, or when the Russians do things in the Ukraine," he explained, "they know the calculus is we are not going to risk World War Three over interests which seem to be in the respective orbs. So they make a calculus.

"And they understand us pretty well," he added. "They are, both the Chinese and the Russians, real students of the United States; and they know what the policy limits are—whether articulated or not. They know what the envelope is."

General Hayden agrees. China and Russia are winning, based on the rules of a dangerous game that they set and they initiated.

"It's a war that advantages them," Michael Hayden told me. "And they have achieved things."

In this final chapter, I will lay out some of their proposed solutions, ranging from hardening US defenses, to increasing deterrence, to the riskier option of undertaking offensive action against China and Russia.

#1: KNOW THE ENEMY

One consistent lesson of the Shadow War is that the United States and the West put themselves in a losing position by succumbing to a fundamental and persistent misreading of Russia and China.

"The hope was they would mirror us," said General Clapper. "When the Soviet era ended, that the Russians would be drawn into the Western capitalist democratic system. Starting with Nixon's visit to China, the great hope was that China would be brought into the Western liberal system."

In fact, decision makers across the public and private sectors are only now abandoning misconceptions of what kind of relationship is possible with Moscow and Beijing.

"It took a long time to sink in, in my personal opinion," said Ashton Carter. "You have a wish for things to have turned out the way we thought they might have in the 1990s.

"With respect to China, some people couldn't see their way past what they imagined to be the economic benefits, either to them or maybe even to the country as a whole, from the economic relationship with China, however exploited it was," Carter said. "They weren't willing to say to themselves, what I would say, which is that China is a communist dictatorship. There's no other way to put it. I'm not suggesting that we try to change its government, but we have to recognize what its government is.

"The same thing is true of Russia," Carter continued. "[It] was basically a fifteen- or twenty-year pattern of aggressively missed signals. It was widespread in the United States and it was also very widespread in Europe."

John Scarlett, who served more than thirty years in British intelligence, including as Moscow station chief, sees a consistent failure to grasp both Russia's and China's motivations.

"In particular, you have to understand the mind of the other side," Scarlett told me. "And we have to ask ourselves, have we understood the mind of the other side?"

His answer is no. To illustrate, he gives the example of the West's reaction to the collapse of the Soviet Union in 1991. In the immediate aftermath and for several years afterward, Western leaders and policy makers assumed that all or most of the Russian leaders and the Russian public suddenly shared Western values and ambitions. Scarlett was posted in Russia at the time, as chief of MI6's Moscow station.

"I don't think we went deep enough into the emotions that had been stirred up by what happened in 1991—the sheer scale of the events," Scarlett told me. "I'm influenced by the fact that I was there at the time: the massive, overnight, unexpected change, the collapse of a superpower almost without warning, the colossal loss of prestige, dislocation at the top national levels."

Those emotions and resentments went far deeper and proved more lasting than Western leaders realized. Years later, those forces would help fuel the rise of Vladimir Putin and Putin's Russia: a declining power bent on recovering its superpower status and the perceived prestige of Soviet times, at the expense of its Western rivals.

"The sense of surprise in 2016 was very striking to our side—the degree to which people were just not ready for it," Scarlett told me. "And that is now changing and that in itself is an improvement."

Today, the collapse of the Soviet Union is a cautionary tale in both Moscow and Beijing. Chinese leaders—as much as their Russian counterparts—study what they perceive as the catastrophic mistakes of 1991 in order to avoid repeating them. The actions of Russia and China today can be explained in part by that fear. They are seeking to grow and retain their power in any way possible—and at the expense of the United States—precisely because they know that such power is fragile. They have seen this movie.

There are of course differences between Russian and Chinese motivations and behaviors. General Clapper notes that the US-Chinese trading relationship—valued at a colossal $600 billion per year—makes both countries dependent on each other to a significant degree. The United States and Russia have no such trading relationship and therefore no such mutual dependence.

"One [distinction] that moderates Chinese behavior to an extent, is the fact that our two economies are inexplicably bound," Clapper notes. "Whereas, with the Russians, it's mutually exclusive. They were during the Soviet era, and they are now."

Winning the Shadow War will require US and Western leaders to understand and acknowledge these motivations—and abandon the misconceptions that had endured for nearly a generation. This change is already taking place in US and Western intelligence and military circles—and that change is reflected in military deployments undertaken by the West. NATO is deploying troops in greater numbers to defend its eastern frontier from Russian military aggression. And in Asia, the United States is displaying its military power, for instance, by increasing freedom-of-navigation operations around China's man-made islands in the South China Sea. (I'll discuss these military moves in greater detail later in this chapter.) Western intelligence agencies are devoting more resources to both human and electronic collection on Russian and Chinese intelligence targets. However, that urgency is missing where it matters most—at the level of the US president and some of his closest advisors and supporters. Without a unified perception of the enemy, a unified response is impossible.

#2: SET RED LINES

Just as the United States and the West often fail to understand the Russian and Chinese mind-set, Moscow and Beijing often misinterpret US and Western intentions. And the West shares responsibility for this. Sending clear signals—and setting clear red lines—is essential to deterring further aggression. For instance, there is broad agreement across the US national security community that the United States has yet to establish a clear red line on election interference, principally due to the president's refusal to prioritize the threat. The United States and NATO have, at times, been clearer and more direct regarding Russian military aggression on the ground in Europe, especially as it relates to the NATO member states closest to the Russian threat: the Baltic nations of Estonia, Lithuania, and Latvia.

"The red line is around the Baltics," said John Scarlett.

Russia has certainly tested this red line, beginning with its bold cyberattack in 2007 and continuing to the present via both cyber intrusion and military activity on the Baltic states' borders. Following Russia's 2014 invasion of Crimea and Eastern Ukraine, NATO has responded with increased military deployments and exercises on its eastern front. The intended message: NATO will not tolerate "little green men" or any other military activity inside the territory of NATO members states. Scarlett believes that is one message, at least, that Moscow has received.

"When you look at their behavior, I don't see signs of an illusion there. I mean they're always going to push their luck," said Scarlett. "And you'll get pushing at the edges, air intrusion, naval activities. That's not the same as real aggression."

The United States has taken a step toward setting such red lines in the cyber realm with US Cyber Command enabled in 2018 to carry out offensive operations to defend US networks. However, as America's seniormost intelligence officials testified in February 2018, President Trump had not yet directed US intelligence agencies to take the measures necessary to repel attacks targeting US elections. With China, while successive administrations have repeatedly warned China to curtail its aggressive cyberattacks on the US private and public sectors, those attacks have continued.

#3: RAISE THE COSTS

Disrupting and deterring hybrid warfare attacks by China and Russia require further raising the costs of their aggression. To date, the United States has stuck to relatively conservative retaliatory measures, including imposing economic costs in the form of sanctions on individuals and entities, filing criminal charges against individuals involved in hostile acts, and publicly naming and shaming states and leaders who order and direct the attacks. These measures have been costly to America's adversaries. For instance, Russia's repeated attempts to lobby President Trump as a candidate and president to remove or weaken the Magnitsky Act reveal its success in penalizing Russian leaders. However, such measures have not measurably changed Russia's aggressive behavior.

"[Vladimir Putin] needs to feel very uncomfortable provoking the United States," said Ash Carter. "I don't think he's been made yet to feel adequately uncomfortable."

Today, national security officials and policy makers recommend more punishing sanctions. These could include sanctions on whole sectors of the Chinese or Russian economies. For instance, the United States could sanction Chinese state-run banks for helping North Korea evade international economic sanctions imposed to curtail its nuclear program. The United States could also impose sanctions on Russian oil exports, as the West did in response to Iran's nuclear program, or Russian banks, by denying or limiting access to US dollar-denominated financial transactions. The latter would severely punish Russian president Vladimir Putin personally. So far, both the Obama and Trump administrations have avoided such broad economic moves.

Knowing the adversary, setting clear red lines for unacceptable behavior, and raising the costs of such unacceptable behavior are steps that US and Western countries have begun to implement in response to Russian and Chinese aggression, if inconsistently. However, US and Western leaders and policy makers are still formulating and debating a more comprehensive response encompassing both defensive and offensive action.

Turning the tide of the Shadow War will require a combination of defense—from defending US space assets against Russian and Chinese space-based weapons to protecting NATO allies in Eastern Europe against Russian military aggression to protecting US election systems from cyberattacks and other foreign interference—and offense, including options ranging from deploying America's own space-based weapons to launching offensive cyberattacks against foreign adversaries.

The challenge with defense is focusing sufficient resources on threats that may not be immediate or clear to US policy makers or the public. The halting response to Russian interference in the 2016 election demonstrates that even a danger that is immediate and clear does not necessarily spark a unified, cogent response.

"We, as a people the United States, don't cope very well with things that haven't happened to us," General Clapper told me. He then laid out a hypothetical to make his point. "If George Tenet, who was then the director of central intelligence, in the summer of 2001, had gone public and said, 'We are very concerned about Al Qaeda. We're onto a plot here involving commandeering aircraft and using them as missiles that crash into buildings. We don't have any specifics. But, as a consequence, what everybody needs to do now is go to the airport two hours early, take your shoes off, only take three ounces of liquids, and subject yourself to an electronic scan and a potential body search.'

"What kind of reaction do you think he would have gotten?" Clapper asked. "He would not have been taken seriously. He would have been laughed offstage, because we can't really get our head around a threat we haven't actually experienced.

"I think that the same is true of the cyber domain and the space domain, because you can't see either one," he said.

The challenge with offense is calibrating offensive responses so as not to inadvertently spark a broader conflict or trigger retaliation that will punish the United States more deeply than any action the United States can mount against its adversaries. The goal is to outplay Russia and China at their own game—that is, to turn the Shadow War against them—while remaining below the threshold of an actual shooting war. But how does the United States strike such a balance? And where is that threshold in cyberspace? On the eastern frontier of NATO? In outer space? Striking that precarious balance is the subject of enormous debate within the national security community today.

Crucially, offense and defense are inextricably intertwined. Just as on the football field, a credible offense is not possible without a credible defense. And, in the Shadow War, it is not clear that the United States and the West can truly mount a credible defense due to the inherent openness of Western societies. This is particularly true in the cyber dimension.

"I've been part of lots of discussions over many long hours in the White House Situation Room about this," General Clapper told me. "The problem that I saw is that it's almost pointless to talk about cyber offense, unless you are very confident in your ability to defend, and then be resilient, if you have a counterretaliation."

The danger is that the United States is so vulnerable in the Shadow War that it may not be able to bear the costs of escalation without real change.

#4: BOLSTER DEFENSES

A. Protect the Home Front

US intelligence and military leaders agree that a winning strategy for the Shadow War begins with a credible defense. And they consistently point to the vulnerability exposed so clearly by Russia's interference in the 2016 election. Russia, China, and other foreign adversaries had attempted for decades to interfere in US elections. However, cyber capabilities vastly increased their ability to do so successfully. This includes both information operations, like the one Russia carried out by stealing and exposing emails from the

Democratic Party and DNC, and more alarming attacks on US election systems themselves. Deputy Attorney General Rod Rosenstein made clear in July 2018 that future ballot fraud by Russia remained a major concern. Russia had not yet carried out attacks to disrupt voting systems, but many officials believe it is just a matter of time. Protecting those systems, therefore, becomes urgent and essential.

"A starting point can be that, where you can, you have to toughen up your defenses," said John Scarlett. "The obvious example of that is defending the electoral process. It takes resources. It takes skill. It takes understanding."

The focus, says Scarlett, must be on maintaining voters' confidence in the electoral system and in electoral results. Once that confidence is lost—and Russia's election interference in 2016 already undermined that confidence for many Americans—it is difficult to repair and restore.

"That's the best way of defeating the attempts to mess about," said Scarlett.

While America's voting system is diverse and decentralized, intelligence and homeland security officials know the targets and have the means to shore up those targets' cyber protections. One obstacle is the states that, by law and tradition, have control over the voting process. Some states have been reluctant to seek and accept federal help, which they perceive as interference.

More broadly, since Russia and China have successfully targeted a whole host of critical US infrastructure—from power plants to the electric grid to water treatment plants and government and private sector email networks and databases—defense must extend across the country. Such efforts have been under way for more than a decade but the attackers have generally been a step ahead. US national security officials are demanding a more urgent national effort to eliminate or at least mitigate their advantage.

A persistent problem is ever-present risk of user error. Russia targeted a whole host of political organizations and individuals by means of blunt cyber tools such as spear phishing (as it did with Clinton campaign chairman John Podesta). With such attacks, all the cyber defense in the world does not make a difference if a user takes the bait.

"This is one case where you're only as strong as your weakest link," Clapper emphasized. "So, that makes it exceedingly difficult for us to do the ironclad defense. I mean, it's almost impossible."

The solution is what cyber experts call "digital hygiene," in which individuals protect themselves, and the system as a whole, by changing habits to avoid aiding cyber attackers. Estonia successfully mobilized its population following Russia's 2007 cyberattack. Today digital hygiene is religion for Estonians. Duplicating such a change in a country like the United States with one hundred times the population is an enormous challenge.

B. Ease Rather than Inflame Internal Divisions

A consistent lesson of the Shadow War is that some of America's wounds are self-inflicted. Some of these are the product of America's open and democratic society, which makes America more vulnerable than its adversaries to Shadow War tactics. For instance, democracies are more vulnerable to information operations because they are simply more open. The Chinese government has its "great firewall" to monitor and limit dissent on the internet. The Russian government has successfully co-opted virtually all of Russian news media.

However, the degree of political division in the United States today exacerbates the country's vulnerability to information operations by making portions of the US public ripe for foreign interference. Russian fake news during the 2016 election found fertile ground in the far corners of the far right, before migrating to larger conservative platforms and, sometimes, right to the US president.

"They grab American-created memes for their social media attacks, generally from the alt-right, occasionally from the president," noted General Hayden. "Covert influence, at its best, can only identify and exploit fractures. And so, to a first order, if we want to defend ourselves against this, it's about us."

In his book *The Assault on Intelligence: American National Security in an Age of Lies*, Hayden tells a revealing story about the explosive growth of conservative outrage at NFL players' "take a knee" protest during the national anthem. Russian bots identified a valuable target for exploitation early on in the controversy and quickly began generating thousands of posts using the hashtags #takeaknee, #NFL, and—in an interesting clue as to the origin of many of these posts—the grammatically incorrect hashtag #taketheknee.

"The most difficult thing to translate are definite articles," Hayden explained with a smile. "'Take the knee' was the third trending hashtag, and then the alt-right picked it up, it goes to Fox, bleeds through Hannity, then from Hannity it goes to *Fox & Friends*, and then he [President Trump] retweets it.

"They all do it for their own purposes, but they all take us to the same place," Hayden said. "To a really important degree, we're our own worst enemies. We give them opportunities."

America will not suddenly heal these divisions. However, some national security experts see ways to make those divisions less easy to exploit, including steps outside the sphere of warfare. Hayden takes aim at America's most dominant social media company,

Facebook, and its algorithms, which he believes help drive Americans more deeply into their respective echo chambers.

"Facebook's business model requires you to stay. The business model—profit—is based on clicks. It's based on time on target," Hayden said. "The longer you stay, the more the algorithm drives you to like-minded people because, scientifically, the algorithm knows you stay longer because you're being reinforced.

"You can complain about Zuckerberg posting fake news. You can complain Zuckerberg should be under the same rules for political advertisement as you are, but it's worse than that," he continued. "It's the very business model that drives us deeper into the darker corners of our own ghettos. It actually is built to divide us as a nation.

"That business, that algorithm, sucks," said Hayden, remarkable words to hear from the former director of the CIA.

His solution? Declare Facebook a utility and allow the government to regulate the social media network, including by mandating changes to its algorithm and business model.

Ashton Carter bristles at the argument that America's vulnerability to foreign information operations is principally a domestic problem.

"We're talking about whether Russia is attacking the United States, not whether that is the lion's share of the problems or issues facing the United States," Carter told me. "Nobody blames Russia for all of the divisions within the United States. That's not my problem with it. My problem with it is that a foreign country is trying to exacerbate things in my country, which is a form of aggression."

Hayden shares the opprobrium at Russia's gall to attack the United States so aggressively and brashly. However, he sees America's growing divisions at home—and the efforts of some politicians to exploit those divisions—as in effect aiding and abetting the enemy.

"There is a big movement afoot, and the president supports it, that's redefining us as a nation of blood, soil, and shared history," said Hayden. "I view that as even more troubling than some other things that are going on, this kind of redefinition of self, and you see it in the immigration policy, you see it in the sense of exclusion, you see it in this transactional view of international relations, rather than relational."

Recognizing those increasing divisions, Russia has found ways to exploit them in order to further divide and weaken the United States.

Healing America's divisions is a broad and long-term political challenge. Better insulating America from fake news and other social media attacks designed to exacerbate those divisions is more attainable in the short term. The United States can look abroad for models. Italy, for instance, has implemented a national program to educate students on how to spot fake news.

In terms of meaningful defense against the Shadow War, many US national security officials believe that educating the US public about these information operations is as important as—or even more important than—any high-tech cyber tools.

C. Enhance Resilience

In both the cyber and space realms, America's technological advancement creates a vulnerability: because the United States is so dependent on space and cyber capabilities, it is therefore more susceptible to attacks targeting those capabilities. Reducing such dependency would create economic and social costs the American public would not tolerate. So today national security officials repeatedly emphasize resilience, that is, building systems that can sustain attacks without shutting down entirely.

In space, this means deploying satellites in greater numbers so that satellite-dependent technologies, such as GPS, can withstand the loss of, or damage to, several satellites without losing functionality on the ground. Satellites are of course very expensive things, both to build and to launch. So the US military and private sectors are designing a new generation of satellites that are smaller and cheaper both to manufacture and send into space. Microsatellites (as discussed in chapter 6) also have potential applications in building resilience in space.

Similarly, in cyber this means designing systems and organizing companies and government institutions so that they can continue to operate in the event of, and even in the midst of, cyberattacks. The goal is to be able to withstand cyberattacks without shutting down entirely, or, ideally, while still performing their core duties.

The 2018 National Defense Strategy emphasized resilience in both the space and cyberspace war-fighting domains: "The Department will prioritize investments in resilience, reconstitution, and operations to assure our space capabilities. We will also invest in cyber defense, resilience, and the continued integration of cyber capabilities into the full spectrum of military operations."[1]

The Defense Department is increasingly demanding such resilience from its many private sector partners and contractors as well.

#5: OFFENSE

When it comes to offensive measures, US strategists do not speak in terms of mobilizing the nation to war against Russia and China. In fact, they emphasize that neither the United States nor Russia or

China wants a shooting war. However, many believe that a credible offensive capability is necessary to deter hostile action short of war.

"In the nuclear era, it was called counterforce and countervalue. Counterforce targets are, 'I'm going to disarm the enemy.' Countervalue is, 'Well, I can't disarm them but I can dissuade them that he ever is going to use them, because I'm going to hold at risk things he holds dear.'"

So how does the United States deploy "countervalue" measures in the Shadow War? One essential question: What do Russia and China hold dear? More pointedly, what do Vladimir Putin and Xi Jinping hold dear? And how can the United States credibly demonstrate its ability—if provoked—to take those things away?

A. Information Ops

One option is for the United States to carry out information operations of its own on Russia and China. Riskiest would be to target Putin and Xi themselves. In the midst of Russia's interference in the 2016 election, the Obama administration considered hacking Putin's vast financial holdings and making them public to expose the extent of his ill-gotten financial assets and undermine his domestic support. To do so during Russia's own presidential election in 2018 would have been particularly impactful.

"What it means is going beyond sanctions, challenging the basic legitimacy of Putin," said Carter.

The United States could also conduct information operations to sow doubts and confusion among the Russian public, much as Russia has done in the United States. Ashton Carter raised the example of exposing the Russian people to the true extent of Russia's military action abroad. For instance, Russia hides how and how many Russian soldiers have lost their lives on the battlefields of

Ukraine and Syria. Relatives are told lies, paid money, and sometimes threatened in order to hide the truth. The United States has the means to prove otherwise—and could spread such information inside Russia.

"We've never really made an effort to explain Russian atrocities in Syria, or Russian body bags coming back from Syria," said Carter. "That is something the United States just hasn't done.

"Making his people wonder what is true and what is not—that is what his trolls try to do to Americans," Carter continued. "That can be done back. Now, that has not been the American tradition. I wouldn't say right [away] we'd be very good at it in general, but I think we have people who can be quite good at it."

Former MI6 chief John Scarlett warns that information operations targeting foreign leaders including Vladimir Putin could backfire.

"Putin is very quick to assume that all sorts of actions are targeted at him, even if they're not," said Scarlett. "And he's highly personalized in his approach, a bit paranoid, really.

"It's why he goes on and on, and has done for years, about those Western attempts to undermine him and undermine Russia and so on, which is a big feature of his thinking and what he says publicly, and I think thinks privately, and the people around him," Scarlett continued.

The danger, he cautions, is that information operations intended to delegitimize Putin could instead legitimize Putin's claims of Western plots to unseat him—sparking further Russian aggression.

"It seems impossible to reassure him at all at this point, so if we were a bit more systematic about it, if it was targeting him, it must just simply provoke him and convince him that his paranoia is correct," warned Scarlett.

To date, US officials insist that the United States does not—and has not since the Cold War—sought to undermine the Russian government by covert means.

"We do speak our minds about Russian human rights, democracy, and so forth. I'm sure Putin doesn't like that, but it's not covert. We conduct cyber espionage, but not cyberattacks on Russia," said Carter.

If the United States were to ratchet up its interference via information operations specifically targeting the Russian or Chinese leaders, the question is how would they respond?

B. Cyberattacks on Infrastructure

At the higher end of the escalation scale would be to conduct cyberattacks on critical infrastructure in Russia, China, and other state actors, or to demonstrate a willingness to do so. In November 2014, a group that identified itself as the "Guardians of Peace" hacked into Sony Pictures, stealing the emails of senior management, salary information, and unreleased copies of films and distributing them across the internet. The United States attributed the attack to North Korea. Its suspected motive: the upcoming release of a film *The Interview*, by Sony Pictures, which painted an unflattering, comical portrait of North Korean dictator Kim Jong-un. The next month, in December 2014, North Korea's internet blacked out for several hours. While the US government never claimed or officially acknowledged involvement, there was widespread speculation that the outage was the result of US retaliation for the Sony hack. And in March 2015, Congressman Michael McCaul, chairman of the House Homeland Security Committee, hinted that was indeed the case. Speaking at an event organized by the Center for Strategic and International Studies, McCaul said, "There were some cyber responses to North Korea."[2]

The far more expansive and consequential cyberattack on Iran's nuclear program via the so-called Stuxnet virus, which was jointly developed and deployed by the United States and Israel, is now seen as a game-changing event in cyber warfare. Discovered in 2010, the Stuxnet attack is believed to have led Russia, China, and other state actors to expand their own offensive cyber capabilities.

Many US national security experts continue to believe that such offensive cyberattacks are justified under some circumstances, including in retaliation for serious and deadly foreign attacks on the United States. The challenge is defining those circumstances and judging the potential consequences. Many caution that such attacks on infrastructure could quickly escalate into a broader cyber war or, in the worst circumstances, a shooting war.

"The thing that has inhibited us is we try to be very precise, surgical, and legalistic," Clapper explained. "You can't count on an adversary to be equally precise, surgical, and legalistic. In the cases where we've contemplated an aggressive counter cyberattack, that always came back to haunt us—the uncertainty about what the adversary you're poking at will do in the way of a counterretaliation."

The United States already has at its disposal the means to conduct such attacks on critical infrastructure abroad. Those capabilities remain secret—and military leaders continue to debate and define the circumstances under which they would recommend their use.

The release of the Pentagon's Nuclear Posture Review in February 2018 raised a new and alarming possibility: that the United States could, under very limited circumstances, order the use of nuclear weapons in response to a devastating cyberattack. US military commanders cautioned that the "nonnuclear strategic attacks" that might precipitate a limited nuclear response were extremely limited. More specifically, they emphasized that cyberattacks, though potentially very damaging, are unlikely to cause the breadth of civilian casualties necessary to justify a nuclear response. However, the public discussion of such a possibility brought the debate on offensive cyber measures to the fore in a way US adversaries could not have missed.

C. Deployments as Deterrence

Since the Shadow War involves the use of hard power, US intelligence and military officials agree that the West must demonstrate its ability and willingness to use hard power as well. To borrow an element of China's "high-low" military strategy, the United States must engage on the low end and be prepared to engage on the high end.

The United States and the West are arguably already pursuing this strategy to some degree with both Russia and China. In the South China Sea, the US Navy's so-called freedom of navigation operations, or FONOPs, are intended to show not only that the United States considers those disputed waters and airspace international, but also that the United States has the ability to project military power there, regardless of Chinese construction or legal claims. Transits by US Navy warships through the Taiwan Straits are designed to send a similar message regarding an independent Taiwan.

US military commanders have communicated even more explicit messages regarding China's man-made islands. In June 2018, a US general made clear that the US military has the ability to destroy the islands swiftly in the event of a military conflict.

"I would just tell you that the United States military has had a lot of experience in the Western Pacific taking down small islands," Lieutenant General Kenneth McKenzie Jr., director of the Joint Staff, told reporters. His comment came in response to a reporter's question about whether the United States has the capability to "blow apart" China's man-made islands.

And he would go on to say that he was merely stating a "historical fact."

"We have a lot of experience, in the Second World War, taking down small islands that are isolated," he explained, adding, "that's a core competency of the U.S. military that we've done before. You shouldn't read anything more into that than a simple statement of historical fact."[3]

US military officials had told me the same in private since China had begun building the islands. And it is true that the United States has powerful missiles capable of making the islands unusable. However, such a public warning was remarkable.

The next day, China fired back at what a high-ranking Chinese general described as "irresponsible" remarks by US officials and military commanders regarding the South China Sea.

"Any irresponsible comments from other countries cannot be accepted," said Lieutenant General He Lei at an international forum in Singapore. "Deploying troops and weapons on islands in the South China Sea is within China's sovereign right to do and allowed by international law."[4]

To send a similar message to Russia regarding threatening military activity along NATO's eastern front, the United States has deployed additional forces to Eastern Europe, including fighter aircraft and a unit of US Marines to the Baltics. NATO has expanded its joint military exercises. And the United States has deployed sub-hunting P-8 Poseidon aircraft to Europe and increased US submarine activity in the region.

"We are all consumed with the soft war, if you will, and we're not paying any attention to what they are doing to build up their capability to fight a hard war," cautioned General Clapper.

"Will they use that? I don't think so, because I do think they respect what would happen to them if they were to do it," he continued. "Particularly if they embarked on a nuclear war. I think they know, just like we know, it would be suicidal."

As the United States has expanded its military deployments and exercises abroad, so have Russia and China. So the challenge becomes matching those deployments and exercises to meet the threat. Finding the right balance is a constant battle with adversaries who are constantly testing the US military's limits.

"As they always have during the Soviet era, they're making a huge investment in something that legitimizes their claim to global great-power status," said Clapper.

D. Weapons in Space?

While the United States has expanded its military footprint down here on earth, US policy makers have yet to decide on the deployment of offensive capabilities in space. As with cyber activity, the fear is that answering offense with offense risks escalating the conflict to a point where both sides lose. In the case of space, even a limited conflict could render large swaths of space unusable for decades. For now the United States is focused on better defending satellites and adding resilience where possible to reduce the effectiveness of any attack on US space assets. But again as with cyber, the question is whether deterrence is possible without a credible threat of retaliation. For now, that question remains unanswered.

#6: WARN OF CONSEQUENCES

Central to effective deterrence is the clearest communication possible of the consequences of aggression. This is as true in the Shadow War as it was during the Cold War, when nuclear conflict was the central risk. However, once again, a clear and consistent message is lacking, particularly from the highest levels of government, even as US military planners devise plans for the Shadow War intended to impose the severest consequences on Russia, China, and other state actors.

"What it means is making more evident—within the limits of secrecy—the full consequences that we are capable of creating for Russia for aggression," said Carter. "Russia has a vast attack surface that it cannot possibly defend. That entire attack surface should be part of the vulnerability that we exploit both in wartime—God forbid it comes to that—or in other instances as a way of showing that aggression is a two-way street.

"Russia is a society with great vulnerability—huge borders, uncontrollably huge borders," Carter continued. "If they act in a place like the Baltics, they are likely to feel pressure from three hundred and sixty degrees."

US leaders must make clear to Russia, China, and other adversaries that the consequences of waging a Shadow War on the United States are clear and devastating.

#7: NEW TREATIES FOR CYBER AND SPACE

To date, Russia and China are fighting an undeclared war on the United States in a conflict with no rules. There is no Geneva

Convention for a Shadow War—no Law of the Sea for space or for cyber. Even during the height of the Cold War, when nuclear war between the superpowers was conceivable, the United States and the Soviet Union, and China to a lesser extent, abided by certain agreements governing the use of force in a conflict. The shared intention was to mitigate the chances that a small conflict escalated into World War III. Those treaties and agreements worked to a degree. There is wide agreement among current and former national security officials that the United States must begin to negotiate with its allies and adversaries to set new rules for new battlegrounds.

"A good analogy for me is the Law of the Sea," General Clapper told me. "[It] is something that has evolved over hundreds of years, and now most seafaring nations understand and abide by the Law of the Sea, generally speaking. And we don't have that.

"Until such times we have internationally recognized norms, like we do for the Law of the Sea," Clapper continued, "it's going to be a Wild West."

#8: MAINTAIN AND STRENGTHEN ALLIANCES

Republicans and Democrats in national security emphasize the importance of maintaining and strengthening international alliances in the face of two nations, China and Russia, whose strategies are based in large part on undermining those alliances to, in effect, divide and conquer the West. In Asia, these include military alliances between the United States and Japan, South Korea, and the Philippines as well as regional organizations such as ASEAN, whose member states have a vested interest in the United States retaining its position as regional power as a counterbalance to a rising China.

In Europe, NATO has not been more relevant or necessary since the collapse of the Soviet Union. European officials, who live closest to the Russian threat, see NATO's role in particularly sharp focus.

"It should suffice, stressing the need for us to act together, for us to promote our common values, to explain our common values, to reassure each other, to secure our defenses. Those are all very powerful arguments," said Scarlett.

Pursuing and defending a rules-based international order has been a central US foreign policy focus for decades through administrations of both parties, as demonstrated by US commitments not just to NATO and other military alliances, but also to the World Trade Organization, the World Bank, the International Monetary Fund, and other international organizations and treaties. Once again, many former national security officials, who served both Republican and Democratic administrations, believe the United States is undermining its own security by undermining such commitments.

"We have our own president challenging it as well," said General Clapper. "Playing into the Russian and Chinese narratives and their hybrid warfare philosophies, when, wittingly or not, the president is aiding and abetting what they are doing."

#9: LEADERSHIP

US intelligence and military leaders emphasize that all of these solutions are impossible without clear leadership from the very top. The United States and the West cannot fight and win the Shadow War, they say, when leaders themselves do not agree on the nature of the adversaries, or even that a Shadow War is taking place.

This inconsistency filters down to the populations of the nations involved.

"It's a question of better explaining and articulating and understanding [the threat] to the wider public," said Scarlett. "So people get used to understanding what's going on, whereas some people now, they're still not really understanding it."

More alarmingly, says Scarlett, "The US system is vulnerable because, really, the behavior of some of the people involved."

President Trump has repeatedly undercut US assessments of Russian malign activities against the United States, most notably, its interference in the 2016 election. He has also at times contradicted US policy positions regarding other Russian acts of aggression, for instance, openly suggesting that Crimea does rightfully belong to Russia. His performance next to Vladimir Putin during their July 2018 summit in Helsinki proved a watershed moment, sparking bipartisan outrage at his failure to confront Putin in person on a whole range of Russian aggression.

Ultimately, the Shadow War may be won or lost on the basis of a sense of shared mission—both within the United States and among Western allies. Division is catnip for Russia and China. In fact, such division is both a product and a goal of the Shadow War. Defeating those tactics requires a unified understanding of what the West is fighting for.

"The best way to defend against this surely is to have a much clearer concept than we do of what we represent, what a liberal democracy actually means," said John Scarlett. "Our leadership ideally should be capable of articulating the values of liberal democracy and in language you can easily understand.

"That is what politicians are for. They're there to represent, articulate, explain, communicate," he added. "This is such an obvious task for them."

Winning the Shadow War requires fighting back with the same degree of commitment and unity demonstrated by US adversaries. And the United States needs to generate that response without a Pearl Harbor or 9/11 moment, precisely because the Shadow War is designed to avoid such moments. The United States, in effect, must turn the most essential feature of the Shadow War on its head. Whether it can successfully do so may determine the state of US national security for years to come.

EPILOGUE

S pend some twenty years as a foreign correspondent covering the world and you begin to notice links between the wars, political upheavals, terror attacks, and other events—often, but not always sad ones—unfolding before your eyes. It's not that news repeats itself. Each story and the people who live those stories are different and deserving of attention in their own right. It's that the root causes and actors behind those events are very much related.

I wrote a piece several years ago titled "The Police State Playbook." It was a reporter's notebook on not one but a series of assignments I had completed in authoritarian states, including Russia and China over many years, but also Myanmar, Zimbabwe, Egypt, Saudi Arabia, and Syria. Each assignment was unique. And yet, in each, I was struck by how countries so different in culture, history, geography, and religion mimicked one another in their exercise of absolute power over their own people.

The "playbook" went something like this. They each blamed any dissent at home on enemies abroad. They each pointed to perceived victimization in the past to rally their people to a common cause today. They each dismissed dissidents and other critics as traitors. They each fed their populations false information. And, together, they justified a whole host of bad to reprehensible behaviors on the raw emotions of fear and hate.

I had witnessed this playbook in action firsthand across

continents. In Myanmar, I was on the ground as the military junta crushed a popular revolution led by thousands of Buddhist monks. In Zimbabwe, I saw Robert Mugabe steal an election from the opposition through a combination of blatant voting fraud and chilling violence, including the suspected murder of the opposition leader's wife. In Egypt, I was there as citizens brought down Hosni Mubarak in popular protests centered in Cairo's Tahrir Square, only to see another general quickly take his place.

Russia and China, however, were the true prodigies of authoritarian power, perfected over decades. I covered a presidential election in Russia that was clearly no election. Opposition candidates and their supporters were followed, harassed, jailed, and sometimes worse. China didn't bother with elections but sought out and stifled the slightest, nascent signs of dissent, quashing opposition before it could make it into the open.

From early on, I often experienced the methods of the Police State Playbook personally. In 1994 in Hong Kong, in my first job as a reporter, the Chinese government ordered the television station I worked for to kill a story I'd reported on the mistreatment of foreign businessmen in mainland China. It was a small story for a small station by a green reporter, but it gave me an early taste of the Chinese government's reach and power. I quit and found a new job.

In 2007 in London, I found myself being tested for exposure to radiation after covering the poisoning of Russian dissident Alexander Litvinenko. I had visited many of the locations he had visited with his Russian assassins, and now, like dozens of other London residents, I was a possible victim of what British officials were calling the first radiological terror attack on international soil.

All these countries—all these police states—had little in common and yet they all exercised power nearly identically, drawing from the same brutal playbook.

Years later, as I returned to the United States and took on new assignments as chief national security correspondent for CNN, I noticed a similar pattern in these police states' exercise of power against their enemies abroad, in particular, the United States. The attacks were disparate—ranging from Russia's stealth invasion of a European country in Ukraine and its interference in the 2016 US election, to China's manufacture of new territory in the South China Sea and aggressive theft of US intellectual property. Over the last several years, their methods became even more expansive and more aggressive, extending this undeclared war from under the waves all the way into space.

However, in the coverage of these events and the public discussion of them in Washington, they were mostly treated as disconnected. I didn't see it that way. Over time, each aggressive action fit into a broader strategy of undermining the United States at every turn and, in the event of a war, leveling the battlefield with the world's biggest military power. China and Russia, divided once again by geography, history, and culture, were following nearly identical strategies to weaken and overtake the United States. It was happening in front of our eyes, and yet the United States did not have a comprehensive strategy to respond. In many circles, US officials and lawmakers didn't recognize there was even a threat to respond to.

Beginning a few years ago, I began to keep a tally of these seemingly disparate events and to make sure that I got myself on the ground—or on the sea, when necessary—to witness and explore them firsthand.

From 2012–13, while working as chief of staff to the US ambassador to China, I witnessed American companies experience the systematic theft of their secrets and intellectual property by the Chinese government and Chinese state-owned enterprises. The theft was not bad behavior by individual actors, but Chinese policy designed to weaken the United States and benefit China. I witnessed other malign Chinese behavior as well, including the silencing of its critics inside China and around the world, even those who had escaped to what they perceived as the safety of the United States. China was feeling empowered to pursue its interests aggressively, unbound by international laws and institutions, or the US defense of those laws and institutions.

When I returned to my work as a journalist, the opportunities to witness this growing conflict expanded. In 2014, I went to Eastern Ukraine at the start of Russia's stealth invasion to see for myself Russian operations intended to destabilize a sovereign European nation. At the time, Ukraine was attempting to hold an election. Russia attempted to undermine the results by burning down polling stations in the east.

In 2015, I received permission from the Pentagon to board a US surveillance jet on an operational mission over the South China Sea—the first time a journalist had been allowed to do so. From the air, I could see for myself how China had, in a matter of months, turned several reefs that had barely peaked above the waves into massive military installations.

Later, I visited the team at the Defense Intelligence Agency that had determined within hours of the loss of MH17 over Ukraine that a Russian missile fired from Russian-controlled territory had taken down the jet and its 298 passengers and crew. What was clear is that the United States knew the very same day that Russia was responsible. What wasn't clear was how the United States would prevent the next act of violence.

Each act of aggression and its aftermath showed both enormous boldness on the part of Moscow and Beijing and plodding equivocation by the United States. And the uncertainty of the US response seemed to fuel the next provocation and power grab. Russia and China clearly had a strategy. The United States and the West did not.

Russia's interference in the 2016 election took its aggression to a new and alarming level. As CNN's chief national security correspondent, I watched and reported as the extent of Russian interference became clear over time, surprising even America's seniormost intelligence officials and hamstringing the Obama administration, which was caught between warning Russia off further interference and avoiding undermining confidence in the election and the expected winner, Hillary Clinton.

Russia's interference exposed America's own vulnerabilities as well. Russian trolls were at times injecting fake news into the political process, at others merely exacerbating existing divisions, over issues ranging from Black Lives Matter to gun violence to climate change. Some of America's most admired companies, such as Facebook, were later found to have been slow to respond and even intent on covering over the true extent of Russian meddling. Two thousand sixteen raised a frightening prospect for Americans: would future elections be truly free and fair and honest?

Now, two years into a new administration, I'm seeing many of the same mistakes repeated. President Trump, in his public comments, refuses to call out Russian aggression and even undermines US intelligence assessments of that aggression. He has taken steps the Obama administration did not, including authorizing offensive cyber measures and highlighting the need for a response to threats to space assets. He has also aggressively called out Chinese theft of US secrets. However, more broadly, the administration, Congress, the defense and intelligence communities, and the private sector have yet to articulate a strategy for responding to the range of Russian and Chinese efforts to undermine the United States. Does the

United States have a plan to win the Shadow War? Does it recognize that a Shadow War is under way?

My personal motivation in writing this book is far from political. I am writing this solely as a concerned American. I've always thought that living overseas cements, rather than weakens, your patriotism. Yes, you can often better identify your country's weaknesses from abroad. But you can also better recognize its strengths. In its vision, there is no question that America has far more to offer the world than China and Russia. The Shadow War is in large part a battle of those visions. I see this book as alerting my fellow Americans to this war and the threat it presents to what our country holds dear.

As the great Eric Sevareid once said of journalists, "All that we try to do is to live at the growing points of society and detect the cutting edges of history."[1]

The Shadow War is one, perhaps defining, cutting edge of American history.

ACKNOWLEDGMENTS

This book rests on the insightful, honest, and sometimes self-critical commentary of some of America's and Europe's senior-most intelligence, military, and political leaders. I am grateful to former DNI James Clapper; former NSA and CIA director Michael Hayden; former Defense Secretary Ashton Carter; former Deputy Defense Secretary Bob Work; former MI6 chief John Scarlett; former NSA deputy director Rick Ledgett; former FBI cyber chief Bob Anderson; General John Hyten, the current head of Strategic Command; the former head of US Air Force Space Command General William Shelton; former National Security advisor Tom Donilon; and former US ambassador to Ukraine (and current ambassador to Greece) Geoffrey Pyatt. In Estonia, President Kersti Kaljulaid, Foreign Minister Sven Mikser, and former defense minister Jaak Aaviksoo generously walked me through their country's sometimes-frightening experience with their giant neighbor to the east. Alexander Hug, until recently of the OSCE, gave me a one-of-a-kind account of the shoot-down of MH17 over Ukraine, which is still one of the most shocking acts of the Shadow War. US Navy captain Ollie Lewis, now of the Joint Staff and formerly commodore of Submarine Squadron 12, gave me valuable insight into submarines' central role in a new "great game" under the waves.

Let me thank CNN for sending me on a series of assignments around the world—from Ukraine to the South China Sea to the

Arctic—and across the US defense and intelligence communities, all of which helped open my eyes to the Shadow War. Jeff Zucker, Rick Davis, and Allison Gollust were supportive from the very beginning, even in the midst of the unending national news cycle. Special thanks to my longtime producer, Jennifer Rizzo, who accompanied me to the front lines of this conflict, including several places where we were not welcome.

Thanks to Gail Ross of Ross Yoon Agency for helping turn a complicated issue into a story worth telling—and thanks to Eric Nelson at HarperCollins for seeing it and backing it wholeheartedly.

On each stop of my global tour of the Shadow War, I encountered US military servicemembers and civil servants who are dedicating their lives to advancing US interests and defending their country against a range of modern-day threats. In the Shadow War, these Americans are just as likely to wear civilian clothes as military uniforms; yet each performs an essential and largely unheralded role in keeping America safe.

I'd like to draw particular attention to a handful of teams that opened their doors to me with generosity and honesty. To the commanders and crews of the USS *Hartford* and USS *Missouri*, thank you for welcoming me aboard. And to all submariners, thank you for sacrificing so much. They call you the "silent service" not just for hiding from adversaries under the waves but for making sacrifices most Americans simply aren't aware of. With my dad having served in the Navy, my membership in the "Order of the Blue Nose" is quite an honor.

Let me thank the US Navy and the VP-45 maritime patrol squadron (the "pelicans") for inviting me and my CNN colleagues on board a P-8 Poseidon surveillance aircraft over the South China Sea—the first time journalists had been allowed on an operational mission of the P-8. Their pilots and flight crews are calm, cool, and collected in an increasingly tense environment.

US Air Force Space Command opened its doors to me and my CNN colleagues at several bases across the country and, in doing so, introduced me to the looming conflict in space and to the "space warriors" preparing for it. They truly live up to their motto, "Guardians of the High Frontier." I'm particularly grateful to the men and women of Schriever Air Force Base and Peterson Air Force Base in Colorado Springs, Colorado; Buckley Air Force Base in Aurora, Colorado; Vandenberg AFB in California; and Offutt AFB in Nebraska, home of US Strategic Command. Let me also thank the highly capable units on the frontlines of this conflict, including the 50th Space Wing at Schriever AFB, known rightfully as the "Masters of Space," and the 460th Space Wing at Buckley AFB.

AGI (Analytical Graphics, Inc.) welcomed us into its sprawling operations center outside Philadelphia and provided analysis and insight throughout this project to keep me up to date on the latest Russian and Chinese activities in space. Special thanks to Paul Graziani, AGI's CEO, and to Bob Hall.

The NSA welcomed my team and me into the agency's Cybersecurity Threat Operations Center (NCTOC), which it describes as the heart of its "24/7/365 cybersecurity operations mission." Inside the NCTOC, that struck me as an accurate description in a battle space where attacks come by the hundreds of thousands every day.

The Defense Intelligence Agency gave my CNN colleagues and me rare access to its Missile & Space Intelligence Center in Huntsville, Alabama. The Huntsville area is seeping with the history of the US space program, with some of its legendary rockets forming an otherworldly skyline. The men and women of the MSIC are the highly capable specialists who determined within hours that it was a Russian rocket fired from Russian-controlled territory that took down Flight MH17 over Ukraine.

Bob Anderson, former executive assistant director of the FBI's

Criminal, Cyber, Response, and Services branch, gave me a candid view of China's continuing theft of US national security secrets and the extent and aggressiveness of its broader efforts to undermine the US.

Andrew Erickson, professor of strategy at the US Naval War College, provided sharp and insightful analysis on China's strategy for the Shadow War in the South China Sea and far beyond. Revealingly, Erickson traces the historical foundations of Beijing's goals all the way back to the founding of Communist China.

Austin Lowe, a China analyst and linguist with a Master of Arts in Asian Studies from Georgetown University's Walsh School of Foreign Affairs and a BA in East Asian Languages and Cultures from Columbia (and who also happens to be my nephew), provided essential analysis and research on China. His time living and studying on the mainland proved invaluable.

CrowdStrike and FireEye provided deep insight and firsthand experience in helping to follow and explain Russian interference in the 2016 election.

My sincere thanks to Julie Tate for her help with fact-checking.

Finally, and most important, this book would not have happened without the support of my family. I am grateful to my wife, Gloria Riviera, who backed this book from concept to the finish, and helped turn some early rough copy into a readable story. Let me also thank our children, Tristan, Caden, and Sinclair. Together, our family already gives me up to the daily demands of covering the news, so surrendering me to writing eighty thousand words on the intricacies of modern warfare was quite a gift. Here's to more of our own adventures around the world in the years to come.

Jim Sciutto
Washington, DC
February 2019

NOTES

CHAPTER 1: INSIDE THE SHADOW WAR

1. "The Litvinenko Inquiry: Report into the Death of Alexander Litvinenko," Chmn, Sir Robert Owen, January 2016, 192.
2. Ibid.
3. "Valery Gerasimov, the General with a Doctrine for Russia," *Financial Times*, September 15, 2017.
4. "The Gerasimov Doctrine: It's Russia's New Chaos Theory of Political Warfare. And It's Probably Being Used on You," Molly McKew, *Politico Magazine*, September/October 2017.

CHAPTER 2: OPENING SALVO (RUSSIA)

1. Estonian Public Radio ("ERR"), April 25, 2017.
2. North Atlantic Treaty Organization, official text, The North Atlantic Treaty, April 4, 1949.
3. Statement by the foreign minister Urmas Paet, *Eesti Paevaleht* (newspaper), May 1, 2007.
4. e-Estonia, Government of Estonia, June 2017.
5. Estonian Defence League ("Kaitseliit").
6. Ibid.

CHAPTER 3: STEALING SECRETS (CHINA)

1. Criminal complaint, *USA v. Su Bin*, 3, filed in US District Court, Central District of California, June 27, 2014.
2. Ibid., 19.
3. Ibid., 20.
4. Ibid., 11.
5. Ibid., 16.
6. Ibid., 17.
7. Ibid., 5.
8. Ibid., 45.
9. Ibid., 15.
10. Ibid., 17.
11. Ibid., 18.
12. Ibid., 17.
13. Ibid., 22.
14. Ibid., 24.
15. Ibid., 35.
16. Statement, US Attorney's Office, Central District of California, August 15, 2014.
17. Statement, US Department of Justice, March 23, 2016.
18. Ibid.

CHAPTER 4: LITTLE GREEN MEN (RUSSIA)

1. "Ukraine Military Plane Shot Down as Fighting Rages," BBC News, July 14, 2014; Aviation Safety Network.
2. At a press conference the next day, a Ukrainian official accused Russia of shooting down an Su-25 fighter jet, initially identifying a Russian military jet as the possible culprit. Pro-Russian rebels on the ground claimed at the time to have shot down two Sukhoi jets.

3. "Helsinki Final Act: 1975–2015," OSCE, 2015.

4. Reports: "Crash of Malaysian Airlines Flight MH17," Dutch Safety Board, September 2014 and October 2015.

5. Ibid.

6. Ibid.

7. Transcript, *Fox News Sunday*, John Kerry interview with Chris Wallace, July 20, 2014.

8. "Kerry: Ukrainian Separatist 'Bragged' on Social Media about Shooting Down Malaysia Flight 17," PolitiFact, July 20, 2014.

9. "A Global Elite Gathering in the Crimea," *Economist*, September 24, 2013.

10. Yalta European Strategy (YES) Conference, Yalta, Crimea, September 2013.

11. "Ukraine's EU Trade Deal Will Be Catastrophic, Says Russia," *Guardian*, September 22, 2013.

12. Putin's Prepared Remarks at 43rd Munich Conference on Security Policy, February 12, 2007.

13. "Putin Hits at US for Triggering Arms Race," *Guardian*, February 10, 2007.

14. Report of the International Advisory Panel (IAP) on its review of the Maidan Investigations, March 31, 2015.

15. Report, Ukraine's Prosecutor General's Office as referenced by IAP investigation.

16. "Ukraine: Excessive Force against Protesters," Human Rights Watch, December 3, 2013.

17. IAP Report, March 31, 2015.

18. Ibid.

19. "Ukraine Crisis: Transcript of Leaked Nuland-Pyatt Call," BBC News, February 7, 2014.

20. Transcript, "The Putin Files," *Frontline*, PBS, June 14, 2017.

21. IAP Report, March 31, 2015.

22. Reuters, March 5, 2014.

23. Transcript, "Address by the President of the Russian Federation," Presidential Executive Office, March 18, 2014.

24. Report: "Flight MH17 Was Shot Down by a BUK Missile from a Farmland Near Pervomaiskyi," Joint Investigation Team (JIT), September 28, 2016.

25. Ibid.

CHAPTER 5: UNSINKABLE AIRCRAFT CARRIERS (CHINA)

1. "Air Force History: The Evacuation of Clark Air Force Base," US Air Force, June 12, 2017.

2. Bill Hayton, *The South China Sea: The Struggle for Power in Asia* (Yale University Press, 2014), 92.

3. Ibid., 97.

4. Stephen Jiang, "Chinese Official: US Has Ulterior Motives over South China Sea," CNN, May 27, 2015.

5. White House transcript, Remarks by President Obama and President Xi of the People's Republic of China in Joint Press Conference, September 25, 2015.

6. "Advance Policy Questions for Admiral Philip Davidson, USN Expected Nominee for Commander, U.S. Pacific Command," Senate Armed Services Committee, April 17, 2018.

7. A notable exception is the Johnson Reef Skirmish of 1988, in which Chinese and Vietnamese naval forces clashed over the Johnson South Reef in the Spratly Islands. The conflict resulted in the deaths of sixty-four Vietnamese soldiers.

8. François-Xavier Bonnet, "Geopolitics of Scarborough Shoal," Research Institute on Contemporary Southeast Asia (IRASEC), November 2012.

9. *The South China Sea Arbitration Award of 12 July 2016*, Permanent Court of Arbitration (PCA Case No 2013-19).

10. State Department Transcript, Daily Press Briefing, Washington, D.C., July 12, 2016.

11. Ian James Storey, "Creeping Assertiveness: China, the Philippines and the South China Sea Dispute," *Contemporary Southeast Asia* 21, no. 1 (April 1999): 96, 99.

12. "Pentagon Says Chinese Vessels Harassed U.S. Ship," CNN, March 9, 2009.

13. "Countering Coercion in Maritime Asia: The Theory and Practice of Gray Zone Deterrence," Center for Strategic and International Studies (CSIS), May 9, 2017.

14. "Advance Policy Questions for Admiral Philip Davidson, USN Expected Nominee for Commander, U.S. Pacific Command," Senate Armed Services Committee, April 17, 2018.

15. Bethlehem Feleke, "China Tests Bombers on South China Sea Island," CNN, May 21, 2018.

CHAPTER 6: WAR IN SPACE (RUSSIA AND CHINA)

1. The Joint Space Operations Center has since been redesignated the Combined Space Operations Center, or CSpOC, as it now includes representation from international partners in the "Five Eyes" intelligence-sharing alliance.

2. Air Force Space Command.

3. P. W. Singer and August Cole, *Ghost Fleet: A Novel of the Next World War* (Boston: Houghton Mifflin Harcourt, 2015).

4. "DepSecDef Work Invokes 'Space Control'; Analysts Fear Space War Escalation," *Breaking Defense*, April 15, 2015.

5. History, 50th Space Wing, Schiever Air Force Base, May 2, 2018.

6. Brig. Gen. David N. Miller Jr. is the Director of Plans, Programs and Financial Management, Headquarters Air Force Space Command, Peterson Air Force Base, Colorado.

7. History of Offutt Air Force Base, United States Air Force, August 2005.

8. "The 50th Anniversary of Starfish Prime: The Nuke That Shook the World," *Discover*, July 9, 2012.

9. "Going Nuclear Over the Pacific," Smithsonian, August 15, 2012.

10. NASASpaceflight.com.

11. Ibid.

CHAPTER 7: HACKING AN ELECTION (RUSSIA)

1. Barbara Starr, "U.S. Official: Spy Plane Flees Russian Jet, Radar; Ends Up over Sweden," CNN, August 4, 2014.

2. "Freed and Defiant, Assange Says Sex Charges 'Tabloid Crap,'" ABC News, December 10, 2010.

3. "A Timeline of the Roger Stone–WikiLeaks Question," *Washington Post*, October 30, 2018.

4. "Putin Says DNC Hack Was a Public Service, Russia Didn't Do It," Bloomberg News, September 2, 2016.

5. "Joint Statement from the Department Of Homeland Security and Office of the Director of National Intelligence on Election Security," ODNI, October 7, 2016.

6. "Transcript: Obama's End-of-Year News Conference on Syria, Russian Hacking and More," *Washington Post*, December 16, 2016.

7. Statement, Office of Senator Jeanne Shaheen, December 12, 2017.

CHAPTER 8: SUBMARINE WARFARE (RUSSIA AND CHINA)

1.   "Sea Ice Tracking Low in Both Hemispheres," National Snow and Ice Data Center, February 6, 2018.

2.   "Russian Mini-Subs Plant Flag at North Pole Sea Bed," *Globe and Mail*, August 2, 2007.

3.   "Is Alaska Next on Russia's List?," *Moscow Times*, October 14, 2014.

4.   "Top Navy Official: Russian Sub Activity Expands to Cold War Level," CNN, April 19, 2016.

5.   "Up to 11 Russian Warships Allowed Simultaneously in Port of Tartus, Syria—New Agreement," RT, January 20, 2017.

6.   NATO Allied Maritime Command.

7.   "Presidential Address to the Federal Assembly," Presidential Executive Office, March 1, 2018.

8.   "Russian Submarines Are Prowling Around Vital Undersea Cables. It's Making NATO Nervous," *Washington Post*, December 22, 2017.

9.   "From This Secret Base, Russian Spy Ships Increase Activity Around Global Data Cables," *Barents Observer*, January 12, 2018.

10.   Ibid.

11.   Ibid.

12.   Ibid.

CHAPTER 9: WINNING THE SHADOW WAR

1.   "2018 National Defense Strategy of the United States of America: Sharpening the American Military's Competitive Edge," Department of Defense, January 2018.

2.   "North Korea Web Outage Response to Sony Hack, Lawmaker Says," Bloomberg News, March 17, 2015.

3. "US Warns of Ability to Take Down Chinese Artificial Islands,"
    CNN, May 31, 2018.

4. "China General He Lei Slams 'Irresponsible Comments' on South
    China Sea," *The Straits Times*, June 2, 2018.

EPILOGUE

1. Eric Sevareid, *Address at Stanford University's 80th Commencement*,
    June 13, 1971.

INDEX

ABOUT THE AUTHOR

Jim Sciutto is CNN's chief national security correspondent and anchor of *CNN Newsroom*. After more than two decades as a foreign correspondent stationed in Asia, Europe, and the Middle East, he returned to Washington to cover the Defense Department, the State Department, and intelligence agencies for CNN. His work has earned him Emmy Awards, the George Polk Award, the Edward R. Murrow Award, and the Merriman Smith Award for excellence in presidential coverage. A graduate of Yale and a Fulbright Fellow, he lives in Washington DC, with his wife, Gloria Riviera, who is a journalist for *ABC News*, and their three children.